MEMOIRS
of the
American Mathematical Society

ISSN 0065-9266 (print)
ISSN 1947-6221 (online)

Number 1228

Continuous-Time Random Walks for the Numerical Solution of Stochastic Differential Equations

Nawaf Bou-Rabee
Eric Vanden-Eijnden

November 2018 • Volume 256 • Number 1228 (fourth of 6 numbers)

AMS AMERICAN MATHEMATICAL SOCIETY

Library of Congress Cataloging-in-Publication Data

Cataloging-in-Publication Data has been applied for by the AMS. See http://www.loc.gov/publish/cip/.

DOI: https://doi.org/10.1090/memo/1228

Memoirs of the American Mathematical Society

This journal is devoted entirely to research in pure and applied mathematics.

Subscription information. Beginning with the January 2010 issue, *Memoirs* is accessible from `www.ams.org/journals`. The 2018 subscription begins with volume 251 and consists of six mailings, each containing one or more numbers. Subscription prices for 2018 are as follows: for paper delivery, US$999 list, US$799.20 institutional member; for electronic delivery, US$879 list, US$703.20 institutional member. Upon request, subscribers to paper delivery of this journal are also entitled to receive electronic delivery. If ordering the paper version, add US$20 for delivery within the United States; US$80 for outside the United States. Subscription renewals are subject to late fees. See `www.ams.org/help-faq` for more journal subscription information. Each number may be ordered separately; *please specify number* when ordering an individual number.

Back number information. For back issues see `www.ams.org/backvols`.

Subscriptions and orders should be addressed to the American Mathematical Society, P. O. Box 845904, Boston, MA 02284-5904 USA. *All orders must be accompanied by payment.* Other correspondence should be addressed to 201 Charles Street, Providence, RI 02904-2213 USA.

Memoirs of the American Mathematical Society (ISSN 0065-9266 (print); 1947-6221 (online)) is published bimonthly (each volume consisting usually of more than one number) by the American Mathematical Society at 201 Charles Street, Providence, RI 02904-2213 USA. Periodicals postage paid at Providence, RI. Postmaster: Send address changes to Memoirs, American Mathematical Society, 201 Charles Street, Providence, RI 02904-2213 USA.

This publication is indexed in *Mathematical Reviews®, Zentralblatt MATH, Science Citation Index®, Science Citation Index™-Expanded, ISI Alerting Services℠, SciSearch®, Research Alert®, CompuMath Citation Index®, Current Contents®/Physical, Chemical & Earth Sciences.*

This publication is archived in *Portico* and *CLOCKSS*.

Printed in the United States of America.

∞ The paper used in this book is acid-free and falls within the guidelines established to ensure permanence and durability.

Visit the AMS home page at `https://www.ams.org/`

10 9 8 7 6 5 4 3 2 1 23 22 21 20 19 18

Contents

Abstract

This paper introduces time-continuous numerical schemes to simulate stochastic differential equations (SDEs) arising in mathematical finance, population dynamics, chemical kinetics, epidemiology, biophysics, and polymeric fluids. These schemes are obtained by spatially discretizing the Kolmogorov equation associated with the SDE in such a way that the resulting semi-discrete equation generates a Markov jump process that can be realized exactly using a Monte Carlo method. In this construction the jump size of the approximation can be bounded uniformly in space, which often guarantees that the schemes are numerically stable for both finite and long time simulation of SDEs. By directly analyzing the infinitesimal generator of the approximation, we prove that the approximation has a sharp stochastic Lyapunov function when applied to an SDE with a drift field that is locally Lipschitz continuous and weakly dissipative. We use this stochastic Lyapunov function to extend a local semimartingale representation of the approximation. This extension makes it possible to quantify the computational cost of the approximation. Using a stochastic representation of the global error, we show that the approximation is (weakly) accurate in representing finite and infinite-time expected values, with an order of accuracy identical to the order of accuracy of the infinitesimal generator of the approximation. The proofs are carried out in the context of both fixed and variable spatial step sizes. Theoretical and numerical studies confirm these statements, and provide evidence that these schemes have several advantages over standard methods based on time-discretization. In particular, they are accurate, eliminate nonphysical moves in simulating SDEs with boundaries (or confined domains), prevent exploding trajectories from occurring when simulating stiff SDEs, and solve first exit problems without time-interpolation errors.

Received by the editor March 11, 2015 and, in revised form, December 4, 2015.

Article electronically published on October 10, 2018.

DOI: https://doi.org/10.1090/memo/1228

2010 *Mathematics Subject Classification.* Primary 65C30; Secondary, 60J25, 60J75.

Key words and phrases. Stochastic differential equations, non-symmetric diffusions, parabolic partial differential equation, monotone operator, discrete maximum principle, Kolmogorov equation, Fokker-Planck equation, invariant measure, Markov jump process, stochastic Lyapunov function, stochastic simulation algorithm, geometric ergodicity.

N. B.-R. was supported in part by National Science Foundation (NSF) grant DMS-1212058. E.V.-E. was supported in part by NSF Materials Research Science and Engineering Center Program Award DMR-1420073; and by NSF Award DMS-1522767.

The first author is affiliated with the Department of Mathematical Sciences, Rutgers University Camden, 311 N 5th Street, Camden, New Jersey 08102. Email: nawaf.bourabee@rutgers.edu.

The second author is affiliated with the Courant Institute of Mathematical Sciences, New York University, 251 Mercer Street, New York, New York 10012-1185. Email: eve2@cims.nyu.edu.

CHAPTER 1

Introduction

1.1. Motivation

Stochastic differential equations (SDEs) are commonly used to model the effects of random fluctuations in applications such as molecular dynamics [**3, 22, 44, 141**], molecular motors [**8**], microfluidics [**100, 146**], atmosphere/ocean sciences [**112**], epidemiology [**115, 139**], population dynamics [**2, 21**], and mathematical finance [**77, 129**]. In most of these applications, the SDEs cannot be solved analytically and numerical methods are used to approximate their solutions. By and large the development of such methods has paralleled what has been done in the context of ordinary differential equations (ODEs) and this work has led to popular schemes such as Euler-Maruyama, Milstein's, etc [**84, 124, 151**]. While the design and analysis of these schemes differ for ODEs and SDEs due to peculiarities of stochastic calculus, the strategy is the same in both cases and relies on discretizing in time the solution of the ODE or the SDE. For ODEs, this is natural: their solution is unique and smooth, and it makes sense to interpolate their path at a given order of accuracy between snapshots taken at discrete times. This objective is what ODE integrators are designed for. The situation is somewhat different for SDEs, however. While their solutions (in a pathwise sense) may seem like an ODE solution when the Wiener process driving them is given in advance, these solutions are continuous but not differentiable in general, and an ensemble of trajectories originates from any initial point once one considers different realizations of the Wiener process.

The simulation of SDEs rather than ODEs presents additional challenges for standard integration schemes based on time-discretization. One is related to the long time stability of schemes, and their capability to capture the invariant distribution of the SDE when it exists. Such questions are usually not addressed in the context of ODEs: indeed, except in the simplest cases, it is extremely difficult to make precise statements about the long time behavior of their solutions since this typically involves addressing very hard questions from dynamical systems theory. The situation is different with SDEs: it is typically simpler to prove their ergodicity with respect to some invariant distribution, and one of the goals of the integrator is often to sample this distribution when it is not available in closed analytical form. This aim, however, is one that is difficult to achieve with standard integrators, because they typically fail to be ergodic even if the underlying SDE is. Indeed the numerical drift in standard schemes can become destabilizing if the drift field is only locally Lipschitz continuous, which is often the case in applications. These destabilizing effects typically also affect the accuracy of the scheme on finite time intervals, making the problem even more severe. Another important difference between SDEs and ODEs is that the solutions of the former can be confined to a certain region of their state space, and it is sometimes necessary to impose boundary conditions at the edge of this domain. Imposing these boundary conditions, or even guaranteeing that the numerical solution of the scheme remains in the correct domain, is typically difficult with standard integration schemes.

In view of these difficulties, it is natural to ask whether one can take a different viewpoint to simulating SDEs, and adopt an approach that is more tailored to the probabilistic structure of their solution to design stable and (weakly) accurate schemes for their simulation that are both provably ergodic when the underlying SDE is and faithful to the geometry of their domain of integration. The aim of this paper is to propose one such strategy. The basic idea is to discretize their solution in space rather than in time via discretization of their infinitesimal generator. Numerous approximations of this second-order partial differential operator are permissible including finite difference or finite volume discretizations [**40, 102, 120, 121, 160**]. As long as this discretization satisfies a realizability condition, namely that the discretized operator be the infinitesimal generator of a Markov jump process on a discrete state space, it defines a process which can be exactly simulated using the Stochastic Simulation Algorithm (SSA) [**40, 47, 102, 160**]. Note that this construction alleviates the curse of dimensionality that limits numerical PDE approaches to low dimensional systems while at the same time permitting to borrow design and analysis tools that have been developed in the numerical PDE context. The new method gets around the issues of standard integrators by permitting the displacement of the approximation to be bounded uniformly and adaptively in space. This feature leads to simple schemes that can provably sample the stationary distribution of general SDEs, and that remain in the domain of definition of the SDE, by construction. In the next section, we describe the main results of the paper and discuss in detail the applications where these types of SDE problems arise.

1.2. Main Results

If the state space of the approximating Markov jump process, or the approximation for short, is finite-dimensional, then a large collection of analysis tools can be transported from numerical PDE theory to assess stability and accuracy of the approximation. For example, the above realizability condition implies that the approximating operator is monotone and that a discrete analog of the maximum principle holds for the approximation [**26**]. However, if the state space of the approximation is not finite-dimensional, then the theoretical framework presented in this paper and described below can be used instead. It is important to underscore that the Monte Carlo method we use to realize this approximation is local in character, and hence, our approach applies to SDE problems which are unreachable by standard methods for numerically solving PDEs. In fact, the approximation does not have to be tailored to a grid. Leveraging this flexibility, in this paper we derive realizable discretizations for SDE problems with general domains using simple finite difference methods such as upwinded and central difference schemes [**52, 103, 148**], though we stress that other types of discretizations that satisfy the realizablility condition also fit our framework.

Since an outcome of this construction is the infinitesimal generator of the approximation, the proposed methods have a transparent probabilistic structure that make them straightforward to analyze using probabilistic tools. We carry out this analysis in the context of an SDE with unbounded coefficients and an unbounded domain. Our structural assumptions allow the drift field of the SDE to be locally Lipschitz continuous, but require the drift field to satisfy weak dissipativity, one-sided Lipschitz continuity and maximal/minimal polynomial growth conditions. Our definition of stability is that the approximation has a stochastic Lyapunov function whenever the underlying SDE has one. This definition is natural when one deals with SDEs with locally Lipschitz drift fields. Indeed, global existence and uniqueness theorems for such SDEs typically rely on the existence of a stochastic Lyapunov function [**78**, **114**].

In this context, here are the main results of the paper regarding realizable discretizations with gridded state spaces. We emphasize that we do not assume that the set of all grid points is finite.

(1) We use Harris Theorem on countable state spaces to prove that the approximation has a unique invariant probability measure and converges exponentially fast to it [**56**, **64**, **123**]. Our main interest in geometric ergodicity is that it typically implies a central limit theorem holds for time averages. To invoke Harris theorem, we prove that the approximation preserves a sharp stochastic Lyapunov function from the true dynamics. Sharp, here, means that this stochastic Lyapunov function dominates the exponential of the magnitude of the drift field, and can therefore be used to prove that all moments of the approximation are bounded.

(2) In addition to playing an important role in proving existence of an invariant probability measure and geometric ergodicity, this stochastic Lyapunov function also helps solve a martingale problem associated to the approximation [**41**, **87**, **133**]. This result justifies a global semimartingale representation of the approximation, and implies that Dynkin's formula holds for conditional expectations of observables that satisfy a mild growth condition. We apply this semimartingale representation to analyze the cost of the approximation. Specifically, we obtain an estimate for the average number of steps the approximation takes on a finite-time interval in terms of the spatial step size parameter.

(3) Applying ideas in [**34**], we derive a stochastic representation of the global error of the approximation. A related formula is used to derive the Talay-Tubaro expansion of the global error of a discrete-time SDE method [**153**]. An analysis of this formula reveals that the order of accuracy of the approximation in representing finite-time conditional expectations and equilibrium expectations is given by the order of accuracy of its generator. Note that geometric ergodicity and finite-time accuracy imply that the approximation is accurate in approximating long-time dynamics, such as equilibrium correlation functions [**19**].

This paper also expands on these results by considering infinitesimal generators that use variable spatial step sizes, and whose state space may not be confined to a grid. In these cases, we embed the 'gridless' state space of the approximation into $\mathbb{R}^n$, and the main issues are:

- the semigroup of the approximation may not be irreducible with respect to the standard topology on $\mathbb{R}^n$; and
- the semigroup of the approximation may not satisfy a Feller property because its infinitesimal generator is an unbounded operator and the sample paths of its associated process are discontinuous in space [**123**, **130**].

To deal with this lack of irreducibility and regularity, we use a weight function to mollify the spatial discretization so that the approximate infinitesimal generator is a bounded linear operator. This mollification needs to be done carefully in order for the approximation to preserve a sharp stochastic Lyapunov function from the true dynamics. In short, if the effect of the mollification is too strong, then the approximation does not have a sharp stochastic Lyapunov function; however, if the mollification is too weak, then the operator associated to the approximation is unbounded.

The semigroup associated to this mollified generator is Feller, but not strongly Feller, because the underlying noise does not have a regularizing effect. However, since the process is Feller and has a stochastic Lyapunov function, we are able to invoke the Krylov-Bogolyubov Theorem to obtain existence of an invariant probability measure for the approximation [**130**]. This existence of an invariant probability measure is sufficient to prove accuracy of the approximation with respect to equilibrium expectations. To summarize, here are the main results of the paper regarding realizable discretizations with gridless state spaces.

(1) The approximation is stable in the sense that it has a sharp stochastic Lyapunov function, and accurate in the sense that the global error of the approximation is identical to the accuracy of its generator. The proof of these properties involves analyzing the action of the generator on certain test functions, which is straightforward to do.
(2) A semimartingale representation for suitable functions of the approximation is proven to hold globally. This representation implies Dynkin's formula holds for this class of functions.
(3) The approximation has an invariant probability measure, which may not be unique. This existence is sufficient to prove that equilibrium expectations of the SDE solution are accurately represented by the approximation.

To be clear, the paper does not prove or disprove uniqueness of an invariant probability measure of the approximation on a gridless state space. One way to perhaps approach this open question is to show that the semigroup associated to the approximation is asymptotically strong Feller [**54**].

Besides these theoretical results, below we also apply our approach to a variety of examples to show that our approach is not only practical, but also permits to alleviate several issues that commonly arise in applications and impede existing schemes.

(1) *Stiffness problems.* Stiffness in SDEs may cause a standard integrator to artificially "explode." For a precise statement and proof of this divergence in the context of forward Euler applied to a general SDE see [**70**]. This numerical instability is a well-known issue with explicit discretizations of nonlinear SDEs [**63, 64, 71, 125, 152**], and it is especially acute if the SDE is stiff, e.g. due to coefficients with limited regularity, which is typically the case in applications. Intuitively speaking, this instability is due to explicit integrators being only *conditionally stable*, i.e. for any fixed time-step they are numerically stable if and only if the numerical solution remains within a sufficiently large compact set. Since noise can drive the numerical solution outside of this compact set, exploding trajectories occur almost surely. The proposed approximation can be constructed to be unconditionally stable, since its jump size can be bounded uniformly in

space. As a result of this bound on the jump size, the mean time lag between successive jumps adapts to the stiffness of the drift coefficients.

(2) *Long-time issues.* SDEs are often used to sample their stationary distribution, which is generically not known, meaning there is no explicit formula for its associated density even up to a normalization constant. Examples include polymer dynamics in extensional, shear, and mixed planar flows [**68, 69, 73, 101, 142, 143**]; polymers in confined flows [**25, 31, 51, 59–62, 74–76, 161, 162, 164**]; polymer conformational transitions in electrophoretic flows [**65, 66, 80, 81, 132, 154**]; polymer translocation through a nanopore [**4, 67, 105, 108, 118**]; and molecular systems with multiple heat baths at different temperatures [**1, 38, 39, 55, 104, 116, 134**] as in temperature accelerated molecular dynamics [**1, 116, 157**]. Unknown stationary distributions also arise in a variety of diffusions with nongradient drift fields like those used in atmosphere/ocean dynamics or epidemiology. The stationary distribution in these situations is implicitly defined as the steady-state solution to a PDE, which cannot be solved in practice. Long-time simulations are therefore needed to sample from the stationary distribution of these SDEs to compute quantities such as the equilibrium concentration of a species in population dynamics, or the equilibrium radius of gyration in Brownian dynamics. This however requires to use numerical schemes that are ergodic when the underlying SDEs they approximate are, which is by no means guaranteed. We show that the proposed approximation is long-time stable as a consequence of preserving a stochastic Lyapunov function from the true dynamics. In fact, this stochastic Lyapunov function can be used to prove that the approximation on a gridded state space is geometrically ergodic with respect to a unique invariant probability measure nearby the stationary distribution of the SDE, as we do in this paper.

(3) *Influence of boundaries.* SDEs with boundaries arise in chemical kinetics, population dynamics, epidemiology, and mathematical finance, where the solution (e.g., population densities or interest rates) must always be nonnegative [**2, 5, 21, 29, 30, 47, 106, 139**]. Boundaries also arise in Brownian dynamics simulations. For example, in bead-spring chain models upper bounds are imposed on bond lengths in order to model the experimentally observed extension of semi-flexible polymers [**127**]. In this context standard integrators may produce updated states that are not within the domain of the SDE problem, which we refer to as nonphysical moves. The proposed approximation can eliminate such nonphysical moves by disallowing jumps outside the domain of definition of the SDE.

(4) *First Exit Problems.* It is known that standard integrators do not work well in simulating first exits from a given region because, simply put, they fail to account for the probability that the system exits the region in between each discrete time step [**48–50**]. Building this capability into time integrators is nontrivial, e.g., one technique requires locally solving a boundary value problem per step, which is prohibitive to do in many dimensions [**113**]. By using a grid whose boundary conforms to the boundary of the first exit problem, the proposed approximation does not make such time interpolation errors.

In addition, the numerical tests illustrate other interesting properties of realizable discretizations such as their accuracy in approximating the spectrum of the infinitesimal generator of various non-symmetric diffusions.

1.3. Relation to Other Works

The class of numerical solutions discussed in the present paper are generalizations of Markov Chain Approximation Methods (MCAM). Although the beginnings of MCAM are evident in the influential work of Aleksandr Khinchin in 1933 [**79**], it was not until the 1970s that Harold Kushner invented MCAM in order to approximate optimally controlled diffusion processes [**88**–**98**]. For overdamped Langevin equations, MCAM combined with SSA has also been developed, see [**40, 102, 120, 121, 160**], and to be specific, this combined approach seems to go back to at least [**40**]. Among these papers, the most general and geometrically flexible realizable discretizations seem to be the finite volume methods developed in [**102**]. To generalize MCAM to diffusion processes with multiplicative noise, we construct realizable discretizations on gridless state spaces (a concept we introduce in this paper), and to ensure scalability, we use SSA to implement our approximation.

Our approach is related to multiscale methods for developing simplified, coarse-grained models for SDEs that leverage the spectrum of suitably defined stochastic matrices [**27, 28, 102, 120, 121, 121, 126, 144, 145**]. However, to be clear, this paper is primarily about approximating the SDE solution, not coarse-graining the dynamics of the SDE.

Our approach also seems to be foreshadowed by Metropolis integrators for SDEs, in the sense that the proportion of time a Metropolis integrator spends in a given region of state space is set by the equilibrium distribution of the SDE, not the time step size [**12, 14, 15, 17**–**19, 137**]. The idea behind Metropolis integrators is to combine an explicit time integrator for the SDE with a Monte Carlo method such that the resulting scheme is ergodic with respect to the stationary distribution of the SDE and finite-time accurate [**18**]. However, a drawback of these methods is that they are typically limited to diffusions that are symmetric [**14**]. In contrast, the methods presented in this paper do not require this structure. Moreover, Metropolis integrators are typically ergodic with respect to the stationary distribution of the SDE, but they may not be geometrically ergodic when the SDE is [**15, 18, 137**]. Intuitively speaking, the reason Metropolis integrators lack geometric ergodicity is that they base their proposal moves on conditionally stable integrators for the underlying SDE. As expected, outside the region of stability of these integrators the Metropolis rejection rate can get arbitrarily close to one. Thus, in these unstable regions, Metropolis integrators may get stuck, and as a consequence, typically do not have a stochastic Lyapunov function. In contrast, the proposed approximations do have a stochastic Lyapunov function for a large class of diffusion processes.

Our approach is also related to, but different from, variable time step integrators for SDEs [**46, 99, 150**]. These integrators are not all adapted to the filtration given by the noise driving the approximation. Let us comment on the adapted (or non-anticipating) technique proposed in [**99**], which is based on a simple and elegant design principle. The technique has the nice feature that it does not require iterated stochastic integrals. The method adjusts its time step size at each step according to an estimate of the 'deterministic accuracy' of the integrator, that is, how well the scheme captures the dynamics in the zero noise limit. The time step size is reduced until this estimate is below a certain tolerance. The aim of this method is to achieve stability by being deterministically accurate. In fact, as a consequence of this accuracy, the approximation 'inherits' a stochastic Lyapunov function from the SDE and is provably geometrically ergodic. Ref. [**99**] also addresses what happens if the noise is large compared to the drift. The difference between these variable time step integrators and the proposed approximation is that the latter obtains its mean time step according to a realizable discretization and attains stability by bounding its jump size.

1.4. Organization of Paper

The remainder of this paper is structured as follows.

Chapter 2: details our strategy to spatially discretize an SDE problem.

- §2.1 defines the term realizable discretization.
- §2.2 defines gridded and gridless state spaces.
- §2.3 derives realizable discretizations in one dimension using finite difference, finite volume, and a probabilistic discretization.
- §2.4 derives realizable discretizations in two dimension using upwinded finite differences for mixed partial derivatives.
- §2.5 presents realizable discretizations in n-dimension, which are based on weighted central difference approximations on a gridless state space.
- §2.6 analyzes the computational cost of our approximation with respect to the dimension of the SDE problem.
- §2.7 generalizes realizable discretizations in n-dimension to diffusion matrices that can be written as a sum of rank one matrices.
- §2.8 proves that if the diffusion matrix is diagonally dominant, then a second-order accurate, spatial discretization is realizable on a gridded state space.

Chapter 3: applies realizable discretizations to SDE problems.

- §3.2 (resp. §3.5) provides numerical evidence that the approximation is stable (i.e. geometrically ergodic) and accurate in representing moments at finite times, mean first passage times, exit probabilities, and the stationary distribution of a one-dimensional process with additive (resp. multiplicative) noise.
- §3.3 provides an asymptotic analysis of the mean-holding time of the approximation, which is used to validate some numerical results.
- §3.6 shows that the approximation eliminates nonphysical moves in the Cox-Ingersoll-Ross model from mathematical finance, is accurate in representing the stationary distribution of this process, and can efficiently handle singularities in SDE problems by using adaptive mesh refinement.
- §3.7 illustrates that the approximation accurately represents the spectrum of the generator in a variety of non-symmetric diffusion processes with additive noise including SDE problems with nonlinear drift fields, internal discontinuities, and singular drift fields.
- §3.9 shows that the approximation on a gridded state space (with variable step sizes) is accurate in representing the stationary distribution of a two-dimensional process with multiplicative noise.
- §3.10 applies the approximation to a Lotka-Volterra process from population dynamics, which shows that the approximation eliminates nonphysical moves, is suitable for long-time simulation, and can capture extinction.

- §3.11 tests the approximation on a Brownian dynamics simulation of a colloidal cluster immersed in an unbounded solvent, and shows that the jump size can be set according to a characteristic length scale in an SDE problem, namely the Lennard-Jones radius.

Chapter 4: analyzes a realizable discretization on a gridded state space.
- §4.1 describes the theoretical context, and in particular, hypotheses on the SDE problem are given in Assumption 4.1.
- §4.2 derives a sharp stochastic Lyapunov function for the SDE solution and the approximation, and uses this Lyapunov function to show that both the SDE solution and the approximation are geometrically ergodic with respect to a unique, invariant probability measure.
- §4.3 solves a martingale problem associated to the approximation, justifies a Dynkins formula for the approximation, and applies this solution to analyze basic properties of realizations of the approximation.
- §4.5 proves that the approximation is accurate with respect to finite-time conditional expectations and equilibrium expectations by using a stochastic representation of the global error [**34**].

Chapter 5: analyzes a realizable discretization on a gridless state space.
- §5.1 presents a mollified, realizable discretization that uses variable step sizes.
- §5.2 proves that this discretization is accurate with respect to the infinitesimal generator of the SDE.
- §5.3 shows that the approximation has a sharp stochastic Lyapunov function provided that the mollification is not too strong.
- §5.4 proves that the approximation is accurate with respect to finite-time conditional expectations.
- §5.5 shows that this realization discretization defines a bounded operator with respect to bounded grid functions and induces a Feller semigroup.
- §5.6 proves that an invariant probability measure exists for the approximation, and uses this existence to prove that the approximation is accurate with respect to equilibrium expectations.

Chapter 6: analyzes tridiagonal, realizable discretizations on gridded state spaces.
- §6.1 derives an explicit formula for the invariant density of a realizable discretization.
- §6.2 quantifies the second-order accuracy of the stationary density of a specific realizable discretization.
- §6.3 derives an explicit formula for the exit probability (or committor function) of a realizable discretization.
- §6.4 derives an explicit formula for the mean first passage time of a realizable discretization.

Chapter 7: recaps the main points of the paper and discusses some directions for future work.

1.5. Acknowledgements

We thank Aleksandar Donev, Weinan E, Harold Kushner, Denis Talay, Raúl Tempone and Jonathan Weare for useful discussions. We also thank Paul Dupuis for bringing Ref. [**34**] to our attention, and specifically, the stochastic representation we use in the proofs of Theorems 4.5.1 and 5.4.1.

CHAPTER 2

Algorithms

2.1. Realizability Condition

Here we present a time-continuous approach to numerically solve an n-dimensional SDE with domain $\Omega \subset \mathbb{R}^n$. Let $f : \Omega \to \mathbb{R}$ be a twice differentiable function. The starting point of our approach is the infinitesimal generator:

$$Lf(x) = Df(x)^T \mu(x) + \mathrm{trace}(D^2 f(x)\sigma(x)\sigma(x)^T) \tag{2.1.1}$$

associated to the stochastic process Y that satisfies the SDE:

$$dY = \mu(Y)dt + \sqrt{2}\sigma(Y)dW \ , \quad Y(0) \in \Omega \tag{2.1.2}$$

where W is an n-dimensional Brownian motion and $\mu : \Omega \to \mathbb{R}^n$ and $\sigma : \Omega \to \mathbb{R}^{n\times n}$ are the drift and noise fields, respectively. Let L^* be the (formal) adjoint operator of L. Given a point mass initial condition $p(0,x) = \delta(x - x_0)$, the Fokker-Planck equation:

$$\frac{\partial p}{\partial t}(t,x) = L^* p(t,x) \tag{2.1.3}$$

describes the evolution of the transition density of $Y(t)$ given that $Y(0) = x_0$. Dual to (2.1.3) is the Kolmogorov equation, which describes the time evolution of a conditional expectation of Y:

$$\frac{\partial u}{\partial t}(t,x) = Lu(t,x) \tag{2.1.4}$$

with initial condition $u(0,x) = f(x)$. The solution of (2.1.4) admits a stochastic representation: $u(t,x) = \mathbb{E}_x f(Y(t))$, where $\mathbb{E}_x$ denotes expectation over $Y(t)$ conditional on $Y(0) = x$. Thus, a weak solution to the SDE (2.1.2) can be obtained by solving (2.1.3) or (2.1.4). In this paper we consider solvers for (2.1.2) that exploit this connection.

The basic idea is to use numerical PDE theory to construct an infinitesimal generator for the approximation. We do this by approximating the infinitesimal generator L of the SDE by a discrete-space generator Q of the form:

$$Qf(x) = \sum_{i=1}^{K} q(x, y_i(x)) \Big(f(y_i(x)) - f(x) \Big) \tag{2.1.5}$$

where we have introduced a reaction (or jump) rate function $q : \Omega \times \Omega \to [0,\infty)$, and K reaction channels (or jump states): $y_i(x) \in \Omega$ for $1 \le i \le K$ and for every $x \in \Omega$. (This terminology comes from chemical kinetics [**47**].) Let h be a spatial step size parameter and p be a positive parameter, which sets the order of accuracy of the method. We require that the infinitesimal generator Q satisfies

(Q1) pth-order local accuracy:

$$Qf(x) = Lf(x) + O(h^p) \quad \forall\, x \in \Omega$$

(Q2) realizability:

$$q(x, y) \geq 0 \quad \forall\, x, y \in \Omega \quad x \neq y$$

The class of functions in (2.1.5) for which (Q1) holds will be specified below. Condition (Q1) is a basic requirement for any spatial discretization of a PDE. If (Q1) holds, and assuming numerically stability, then the approximation is pth-order accurate on finite-time intervals (see e.g. Theorem 4.5.1 for a precise statement). Condition (Q2) is not a standard requirement, and is related to the probabilistic interpretation of solutions to the Kolmogorov equation. In words, (Q2) states that the weights in the finite differences appearing in (2.1.5) are non-negative. We call a spatial discretization that satisfies (Q2) realizable, because it induces a continuous-time Markov jump process. Given that L itself induces a Markov process, this is a natural requirement to impose on Q. In contrast to L, however, Q induces a process that can be simulated exactly in time using the Stochastic Simulation Algorithm (SSA) [**40, 47, 102, 160**]. This simulation does not require explicitly gridding up any part of state space, since its inputs – including rate functions and reaction channels – can be computed online. Without (Q2), we stress that the approximation would be a standard numerical PDE solver and limited to low-dimensional systems by the curse of dimensionality – more on this point in §2.6. We also stress that the reaction channels in (2.1.5) can be designed so that the spatial displacement (or jump size) $(y_i - x)$ is uniformly bounded for $1 \leq i \leq K$, where as a reminder K is the # of channels.

Let X be the process induced by Q. To emphasize that exactly simulating X is simple, let us briefly recall how SSA works.

ALGORITHM 2.1 (Stochastic Simulation Algorithm [**47**]). Given the current time t, the current state of the process $X(t) = x$, and the K jump rates evaluated at this state $\{q(x, y_i(x))\}_{i=1}^K$, the algorithm outputs an updated state of the system $X(t + \delta t)$ at time $t + \delta t$ in two sub-steps.

(Step 1): obtains δt by generating an exponentially distributed random variable with parameter

$$\lambda(x) = \sum_{i=1}^{K} q(x, y_i(x)).$$

(Step 2): updates the state of the system by assuming that the process moves to state $y_i(x)$ with probability:

$$\mathbb{P}(X(t + \delta t) = y_i(x) \mid X(t) = x) = \frac{q(x, y_i(x))}{\lambda(x)}$$

for $1 \leq i \leq K$.

We emphasize that conditional on $X(t) = x$, the random time δt from (Step 1) is independent of the random spatial update $X(t+\delta t)$ from (Step 2). Algorithm 2.1 suggests the following compact way to write the action of the generator Q in (2.1.5):

$$Qf(x) = \lambda(x)\, \mathbb{E}(f(\xi_x) - f(x)) \tag{2.1.6}$$

where the expectation is over the random variable $\xi_x \in \Omega$, whose state space is the set of reaction channels $\{y_i(x) \mid 1 \leq i \leq K\}$ and whose probability distribution is

$$\mathbb{P}(\xi_x = y_i(x)) = \frac{q(x, y_i(x))}{\lambda(x)}, \quad \text{for } 1 \leq i \leq K.$$

To summarize, this paper proposes to (weakly) approximate the solution of a general SDE using a Markov process that can be simulated using Algorithm 2.1, whose updating procedure entails generating a random time increment in (Step 1) and firing a reaction in (Step 2). By restricting the firings in (Step 2) to reaction channels that are a bounded distance away from the current state x, the approximation is often numerically stable, a point we elaborate on in §4.2. From (Step 1) note that the mean time increment (or mean holding time) is $\lambda(x)^{-1}$. Thus, the spatial discretization of L determines the average amount of time the process spends at any state. Moreover, note that the process satisfies: $X(s) = x$ for $s \in [t, t + \delta t)$. The latter property implies that sample paths of the process are right-continuous in time with left limits (or càdlàg). Also note that the process only changes its state by jumps, and hence, is a pure jump process. Since the time and state updates only depend on the current state $X(t_0)$, and the holding time in (Step 1) is exponentially distributed, this approximation has the Markov property and is adapted to the filtration of the underlying noise. Thus, the approximation is a Markov jump process. As we show in Chapter 4, these properties of the approximation simplify our analysis.

Next we discuss the state space of the approximation, and then construct infinitesimal generators that meet the requirements (Q1) and (Q2).

2.2. Gridded vs Gridless State Spaces

We distinguish between two types of discretization based on the structure of the graph that represents the state space of the approximation. As is standard, we say that a graph is connected if there is a path of finite length joining every pair of vertices in the graph.

DEFINITION 2.2.1. A gridded (resp. gridless) state space consists of points which form a connected (resp. disconnected) graph.

We stress that both state spaces permit capping the spatial step size of the approximation. Moreover, the SSA on either state space only requires local evaluation of the SDE coefficients. Gridded state spaces are appealing theoretically because a generator on such a space can often be represented by a matrix, and in low dimensional problems, the spectral properties of these matrices (e.g. low-lying eigenvalues, leading left eigenfunction, or a spectral decomposition) can be computed using a standard sparse matrix eigensolver. When the state space is unbounded this matrix is infinite. However, this matrix can sometimes still be approximated by a finite-dimensional matrix, and the validity of this approximation can be tested by making sure that numerically computed quantities like the spectrum do not depend on the truncation level. We will return to this point when we consider concrete examples in Chapter 3. In order to be realizable, the matrix Q associated to the linear operator Q on a gridded state space must satisfy the *Q-matrix property*:

$$\mathsf{Q}_{ij} \geq 0 \quad \text{for all } i \neq j \quad \text{and} \quad \sum_j \mathsf{Q}_{ij} = 0. \tag{2.2.1}$$

Indeed, the Q-matrix property is satisfied by the matrix Q if and only if its associated linear operator Q is realizable. Traditional finite difference and finite volume discretizations have been successfully used to construct Q-matrix discretizations for special classes of SDEs: multi-dimensional, overdamped Langevin equations [**40, 102, 120, 121, 160**], and independently, for time-homogeneous SDEs with diagonally dominant diffusion matrices [**88–98**]. However, Q-matrix discretizations on a gridded state space become trickier to design for more general SDEs, particularly when the noise is multiplicative, even in low dimensional problems. This issue motivates designing approximations on gridless state spaces. Let us elaborate on this point by considering realizable discretizations in one, two, and then many dimensions. Our most general realizable discretization is given in §2.7.

A comment on notation: we will use $\tilde{Q}$ (resp. Q), with possible subscripts, to denote an operator on a gridded (resp. gridless) state space, and with a slight abuse of notation, we may use the same symbol to denote an infinitesimal generator and its associated matrix.

2.3. Realizable Discretizations in 1D

In one dimension, SDEs typically possess a symmetry property, which we briefly recall since we expect the proposed schemes to accurately represent this structure. To prove this property, it is sufficient to assume that the function $M(x) = \sigma(x)^2$ is differentiable and satisfies $M(x) > 0$ for all $x \in \Omega = [\alpha, \beta]$, where Ω is the state space of the diffusion and we assume that $-\infty \leq \alpha < \beta \leq \infty$. Under these assumptions the generator (2.1.1) can be written in conservative form as:

$$(Lf)(x) = \frac{1}{\nu(x)}(M(x)\nu(x)f'(x))' \tag{2.3.1}$$

where $\nu(x)$ is given by:

$$\nu(x) = \frac{1}{M(x)} \exp\left(\int_a^x \frac{\mu(s)}{M(s)} ds\right) \tag{2.3.2}$$

for some $a \in \Omega$. Note that $\nu(x)$ is positive for all $x \in \Omega$. Similarly, the SDE (2.1.2) can be written in the form:

$$dY = -M(Y)U'(Y)dt + M'(Y)dt + \sqrt{2}\sigma(Y)dW \tag{2.3.3}$$

where $U(x)$ is the free energy of $\nu(x)$ defined as:

$$U(x) = -\log(\nu(x)).$$

Let $L^2(\mathbb{R}, \nu(x)dx)$ denote the space of functions square integrable with respect to the measure $\nu(x)dx$. Let $\langle \cdot, \cdot \rangle_\nu$ denote the ν-weighted L^2 inner product defined as:

$$\langle f, g \rangle_\nu = \int_\Omega f(x)g(x)\nu(x)dx \quad \forall f, g \in L^2(\Omega, \nu(x)dx).$$

Integration by parts shows that L is symmetric with respect to this inner product i.e.

$$\langle Lf, g \rangle_\nu = \langle f, Lg \rangle_\nu \quad \forall f, g \in C_0^\infty(\Omega). \tag{2.3.4}$$

A similar calculation shows that the measure $\nu(x)dx$ is infinitesimally invariant

$$\int_\Omega Lf(x)\nu(x)dx = 0 \quad \forall f \in C_0^\infty(\Omega)$$

which implies that $L^*\nu(x) = 0$ for all $x \in \Omega$. Furthermore, if the positive function $\nu(x)$ is integrable over Ω, i.e. $Z = \int_\Omega \nu(x)dx < \infty$, and the diffusion corresponding to L is recurrent, then $\nu(x)/Z$ is an invariant probability density (or stationary density) for the diffusion process Y. For more detail regarding these statements including proofs see, e.g., [**128**, §4.8 & §5.1].

We stress that the infinitesimal generator of SDEs in more than one dimension often do not satisfy the symmetry property (2.3.4). To avoid losing generality, we will not assume that L is symmetric. In this context, we present realizable finite difference, finite volume, and probabilistic discretizations of L. As we show in Proposition 6.1.1, all of these infinitesimal generators are symmetric in a weighted inner product space.

2.3.1. Finite Difference. Let $\{x_i\}$ be a collection of grid points over $\Omega \subset \mathbb{R}$ as shown in Figure 2.1. The distance between neighboring points is allowed to be variable. Thus, it helps to define forward, backward and average spatial step sizes:

$$\delta x_i^+ = x_{i+1} - x_i \ , \quad \delta x_i^- = x_i - x_{i-1} \ , \quad \delta x_i = \frac{\delta x_i^+ + \delta x_i^-}{2} \ , \quad \text{resp.} \tag{2.3.5}$$

at every grid point x_i. The simplest way to construct a realizable discretization of L is to use an upwinded and a central finite difference approximation of the first and second-order derivatives in L, respectively. To derive this discretization, it is helpful to write the generator L in (2.3.1) in terms of the functions $M(x)$ and $U(x)$ as follows:

$$(Lf)(x) = -M(x)U'(x)f'(x) + (M(x)f'(x))'. \tag{2.3.6}$$

To obtain a realizable discretization, the first term of (2.3.6) is discretized using a variable step size, first-order upwind method:

$$-M_iU_i'f_i' \approx (-M_iU_i' \vee 0)\frac{f_{i+1} - f_i}{\delta x_i^+} + (-M_iU_i' \wedge 0)\frac{f_i - f_{i-1}}{\delta x_i^-}.$$

Here we employ the shorthand notation: $f_i = f(x_i)$ for all grid points x_i (and likewise for other functions), and

$$a \vee b = \max(a, b) \quad \text{and} \quad a \wedge b = \min(a, b)$$

for any real numbers a and b. By taking the finite difference approximation on the upstream side of the flow, i.e., going backward or forward in the finite difference approximation of f_i' according to the sign of the coefficient $-M_iU_i'$, this discretization is able to simultaneously satisfy first-order accuracy (Q1) and realizability (Q2). A variable step size, central scheme can be used to approximate the second-order term in (2.3.6):

$$(Mf')_i' \approx \frac{1}{2\delta x_i}\left((M_{i+1} + M_i)\frac{f_{i+1} - f_i}{\delta x_i^+} - (M_{i-1} + M_i)\frac{f_i - f_{i-1}}{\delta x_i^-}\right).$$

It is straightforward to verify that this part of the discretization obeys (Q1) with $p = 2$ and since $M(x) > 0$ satisfies (Q2) as well. The resulting discrete generator is a Q-matrix with off-diagonal entries

$$\begin{cases} (\tilde{Q}_u)_{i,i+1} = \dfrac{1}{\delta x_i^+}\left((-M_iU_i' \vee 0) + \dfrac{M_{i+1} + M_i}{2\,\delta x_i}\right) \\ (\tilde{Q}_u)_{i,i-1} = \dfrac{1}{\delta x_i^-}\left(-(-M_iU_i' \wedge 0) + \dfrac{M_{i-1} + M_i}{2\,\delta x_i}\right) \end{cases} \tag{2.3.7}$$

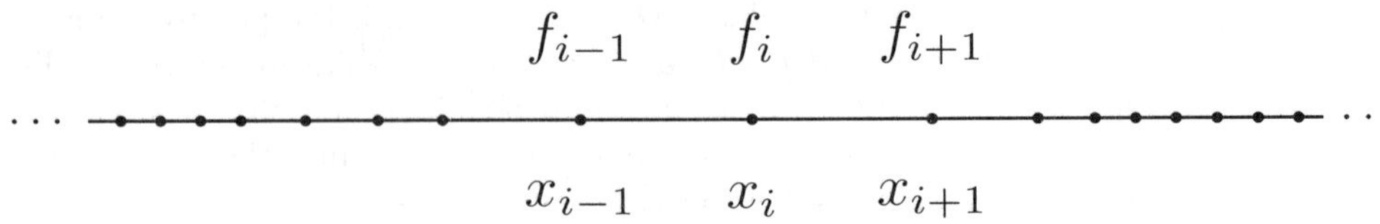

FIGURE 2.1. **Finite Difference Grid.** Consider non-uniformly spaced grid points $\{x_j\}$ and corresponding function values $\{f_j = f(x_j)\}$. This diagram depicts a three-point stencil centered at grid point x_i with corresponding function values. This stencil is used to approximate the generator L using a realizable, finite difference approximation. See (2.3.7) for the entries of the resulting Q-matrix.

Note that to specify a Q-matrix, it suffices to specify its off-diagonal entries and then use (2.2.1) to determine its diagonal entries. Also, note that this scheme does not require computing the derivative of $M(x)$, even though $M'(x)$ appears in the drift coefficient of (2.3.3). The overall accuracy of the scheme is first-order in space. Remarkably, the asymptotic mean holding time of this generator is exact when, roughly speaking, the drift is large compared to the noise. We will prove and numerically validate this statement in §3.3.

2.3.2. Finite Volume. Next we present a finite volume discretization of L. This derivation illustrates the approach taken in [**102**] to construct realizable discretizations for self-adjoint diffusions. For this purpose discretize the region Ω using the partition shown in Figure 2.2. This partition consists of a set of cells $\{\Omega_i\}$ with centers $\{x_i\}$ spaced according to (2.3.5). The cells are defined as $\Omega_i = [x_{i-1/2}, x_{i+1/2}]$ where the midpoint between cell centers $x_{i+1/2} = (x_i + x_{i+1})/2$ gives the location of each cell edge. From (2.3.5) note that the length of each cell is $x_{i+1/2} - x_{i-1/2} = \delta x_i$.

Integrating the Fokker-Planck equation (2.1.3) over cell Ω_i yields:

$$\begin{aligned} \int_{\Omega_i} \frac{\partial p}{\partial t}(t,x)dx &= \int_{\Omega_i} \frac{\partial}{\partial x}\left(\nu(x)M(x)\frac{\partial}{\partial x}\frac{p(t,x)}{\nu(x)}\right)dx \\ &= \left(\nu(x)M(x)\frac{\partial}{\partial x}\frac{p(t,x)}{\nu(x)}\right)\Bigg|_{x_{i-1/2}}^{x_{i+1/2}} \\ &= F_{i+1/2} - F_{i-1/2} \end{aligned}$$

Here we have introduced $F_{i+1/2}$ and $F_{i-1/2}$, which are the probability fluxes at the right and left cell edge, respectively. Approximate the solution $p(t,x)$ to the Fokker-Planck equation by its cell average:

$$\bar{p}_i(t) \approx \frac{1}{\delta x_i}\int_{\Omega_i} p(t,x)dx.$$

In terms of which, approximate the probability flux using

$$F_{i+1/2} \approx \overbrace{\frac{\nu_{i+1} + \nu_i}{2}}^{\approx \nu_{i+1/2}} \overbrace{\frac{M_{i+1} + M_i}{2}}^{\approx M_{i+1/2}} \underbrace{\frac{1}{\delta x_i^+} \left(\frac{\bar{p}_{i+1}(t)}{\nu_{i+1}} - \frac{\bar{p}_i(t)}{\nu_i} \right)}_{\approx \left(\frac{\partial}{\partial x} \frac{p(t,x)}{\nu(x)} \right)_{i+1/2}} . \tag{2.3.8}$$

These approximations lead to a semi-discrete Fokker-Planck equation of the form:

$$\frac{d\bar{p}}{dt}(t) = \tilde{Q}_\nu^\star \bar{p}(t)$$

where $\tilde{Q}_\nu^\star$ is the transpose of the matrix $\tilde{Q}_\nu$, whose off-diagonal entries are

$$\begin{cases} (\tilde{Q}_\nu)_{i,i+1} = \dfrac{1}{\delta x_i \, \delta x_i^+} \dfrac{M_{i+1} + M_i}{2} \dfrac{\nu_{i+1} + \nu_i}{2\nu_i} \\ (\tilde{Q}_\nu)_{i,i-1} = \dfrac{1}{\delta x_i \, \delta x_i^-} \dfrac{M_{i-1} + M_i}{2} \dfrac{\nu_{i-1} + \nu_i}{2\nu_i} \end{cases} \tag{2.3.9}$$

This infinitesimal generator is a second-order accurate approximation of L and is symmetric in the following way. Let $\bar{\nu}$ be the density ν in (2.3.2) weighted by the length of each cell

$$\bar{\nu}_i = \delta x_i \, \nu_i .$$

Using the definition in (2.3.9), and since $\delta x_i^+ = \delta x_{i+1}^-$ by (2.3.5), we have that

$$(\tilde{Q}_\nu)_{i,i+1} \, \bar{\nu}_i = (\tilde{Q}_\nu)_{i+1,i} \, \bar{\nu}_{i+1} .$$

This identity implies that $\tilde{Q}_\nu$ is symmetric in a $\bar{\nu}$ weighted inner product space, see Proposition 6.1.1 for a detailed proof. Other approximations to the probability flux also have this property e.g. the approximation obtained by replacing the midpoint approximation of $\nu_{i+1/2}$ in (2.3.8) by:

$$\nu_{i+1/2} \approx \nu_i \wedge \nu_{i+1} \quad \text{or} \quad \nu_{i+1/2} \approx \exp\left(-\frac{U_{i+1} + U_i}{2} \right) .$$

(As a caveat, note that the former choice reduces the order of accuracy of the discretization because it uses the minimum function.) In fact, the evaluation of U can be bypassed by approximating the forward/backward potential energy differences appearing in (2.3.9) by the potential force to obtain the following Q-matrix approximation:

$$\begin{cases} (\tilde{Q}_c)_{i,i+1} = \dfrac{1}{\delta x_i \, \delta x_i^+} \dfrac{M_{i+1} + M_i}{2} \exp\left(-U_i' \dfrac{\delta x_i^+}{2} \right) \\ (\tilde{Q}_c)_{i,i-1} = \dfrac{1}{\delta x_i \, \delta x_i^-} \dfrac{M_{i-1} + M_i}{2} \exp\left(U_i' \dfrac{\delta x_i^-}{2} \right) \end{cases} \tag{2.3.10}$$

In terms of function evaluations, (2.3.10) is a bit cheaper than (2.3.9), and this difference in cost increases with dimension. Under natural conditions on $M(x)$ and $U'(x)$, we will prove that (2.3.10) possesses a second-order accurate stationary density, see Proposition 6.2.1.

For a more general construction of realizable finite-volume methods for overdamped Langevin equations, we refer to [**102**].

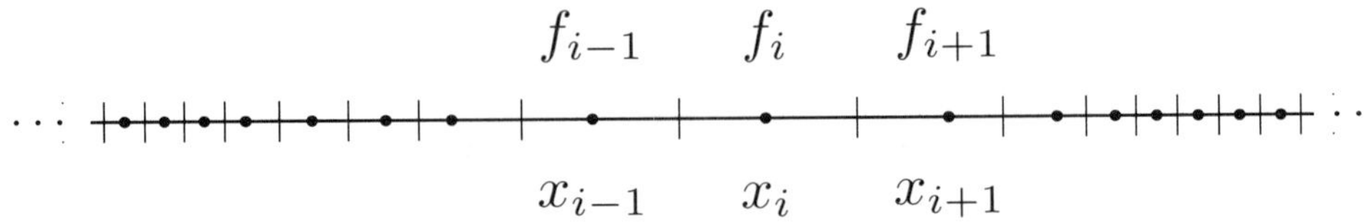

FIGURE 2.2. **Finite Volume Cells.** Consider non-uniformly spaced points $\{x_j\}$ and corresponding function values $\{f_j = f(x_j)\}$. This diagram depicts a set of cells with cell centers given by the points $\{x_j\}$ and with cell edges located in between them at $(x_j+x_k)/2$ for $k = j-1, j+1$. This notation is used to derive a realizable, finite volume discretization of the generator L. See (2.3.9) for the resulting Q-matrix.

2.3.3. Time-Continuous Milestoning. Finally, we consider a 'probabilistic' discretization of the generator L. The idea behind this construction is to determine the jump rates entering (2.1.5) from the exit probability and the mean first passage time (MFPT) of the SDE solution.

For this purpose, let $\tau_a^x = \inf\{t \geq 0 \mid Y(0) = x \ , \ Y(t) = a\}$ denote the first time $Y(t)$ hits a given that $Y(0) = x$; and, define the exit probability (or what is also known as the committor function) to be $q_{a,b}(x) = \mathbb{P}(\tau_b^x < \tau_a^x)$, which gives the probability that Y first hits b before it hits a, starting from any state x between them. This function satisfies the following boundary value problem:

$$Lq_{a,b}(x) = 0 \ , \quad q_{a,b}(a) = 0 \ , \quad q_{a,b}(b) = 1. \tag{2.3.11}$$

From the solution of (2.3.11), the exit probabilities of Y with $Y(0) = x_i$ can be written as

$$\begin{aligned} q_{x_{i-1},x_{i+1}}(x_i) &= \mathbb{P}(\tau_{x_{i+1}}^{x_i} < \tau_{x_{i-1}}^{x_i}) = \frac{q_i - q_{i-1}}{q_{i+1} - q_{i-1}} \\ q_{x_{i+1},x_{i-1}}(x_i) &= \mathbb{P}(\tau_{x_{i-1}}^{x_i} < \tau_{x_{i+1}}^{x_i}) = \frac{q_{i+1} - q_i}{q_{i+1} - q_{i-1}} \end{aligned} \tag{2.3.12}$$

where $q(x)$ is the general solution to $Lq(x) = 0$ for all $x \in \Omega$. We use these expressions below to define the transition probabilities of the approximation from x_i to its nearest neighbors. In addition, recall that the MFPT of Y to $(x_{i-1}, x_{i+1})^c$ given that $Y(0) = x \in [x_{i-1}, x_{i+1}]$ is the unique solution $u_i(\cdot)$ to the following boundary value problem:

$$Lu_i(x) = -1 \ , \quad x_{i-1} < x < x_{i+1} \ , \quad u_i(x_{i-1}) = u_i(x_{i+1}) = 0. \tag{2.3.13}$$

In particular, given that the process starts at the grid point x_i, the MFPT of $Y(t)$ to $(x_{i-1}, x_{i+1})^c$ is given by

$$\mathbb{E}(\tau_{x_{i-1}}^{x_i} \wedge \tau_{x_{i+1}}^{x_i}) = u_i(x_i).$$

In terms of these solutions, we construct an infinitesimal generator for the approximation by specifying its mean holding time in the current state and its transition probabilities to nearest neighbors. Since the process Y initiated at x_i spends on average $u_i(x_i)$ time units before it first hits a nearest neighbor of x_i, the mean holding time of the approximation in state x_i is set equal to:

$$((Q_e)_{i,i+1} + (Q_e)_{i,i-1})^{-1} = u_i(x_i). \tag{2.3.14}$$

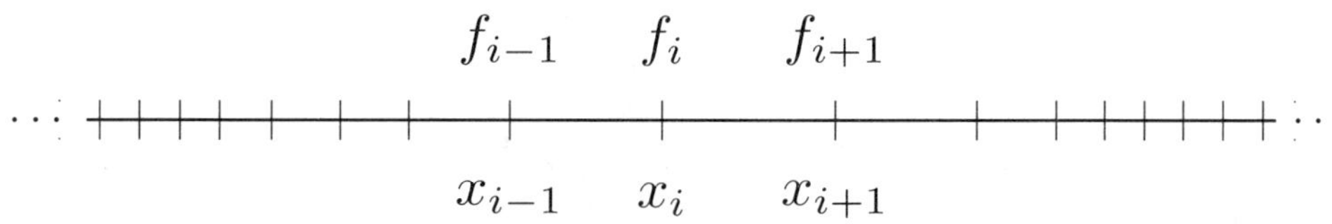

FIGURE 2.3. **Milestones.** Consider a non-uniform (bounded or unbounded) grid, a set of milestones $\{x_j\}$ on this grid, and corresponding function values $\{f_j = f(x_j)\}$ as shown in this diagram. Using this notation a realizable discretization which retains the exact transition probabilities and mean holding times of the SDE is given in (2.3.16).

Similarly, from the solution of (2.3.12), the transition probabilities of the approximation to x_{i+1} and x_{i-1} are resp. set equal to:

$$\begin{aligned} \frac{(Q_e)_{i,i+1}}{(Q_e)_{i,i+1} + (Q_e)_{i,i-1}} &= \frac{q_i - q_{i-1}}{q_{i+1} - q_{i-1}} \\ \frac{(Q_e)_{i,i-1}}{(Q_e)_{i,i+1} + (Q_e)_{i,i-1}} &= \frac{q_{i+1} - q_i}{q_{i+1} - q_{i-1}} \end{aligned} \tag{2.3.15}$$

Incidentally, these are the transition probabilities used in discrete-time optimal milestoning [**158**], where the milestones correspond to the grid points depicted in Figure 2.3. Combining (2.3.14) and (2.3.15) yields:

$$\begin{cases} (Q_e)_{i,i+1} = \dfrac{q_{x_{i-1},x_{i+1}}(x_i)}{u_i(x_i)} = \dfrac{q_i - q_{i-1}}{q_{i+1} - q_{i-1}} \dfrac{1}{u_i(x_i)} \\ (Q_e)_{i,i-1} = \dfrac{q_{x_{i+1},x_{i-1}}(x_i)}{u_i(x_i)} = \dfrac{q_{i+1} - q_i}{q_{i+1} - q_{i-1}} \dfrac{1}{u_i(x_i)} \end{cases} \tag{2.3.16}$$

which shows that a Q-matrix can be constructed in one-dimension that uses the exact exit probability and MFPT of Y. We emphasize that this discretization is still an approximation of the infinitesimal generator of Y, and hence, does not exactly reproduce other statistics of Y, as pointed out in [**158**]. We will return to this point in §6.4 where we analyze some properties of this discretization. In §3.3, we will apply this discretization to validate an asymptotic analysis of the mean holding time of the previously introduced realizable discretizations in one dimension.

To recap, the construction of Q_e uses the exit probability (or committor function) of Y to compute the transition probabilities, and the MFPT of Y to compute the mean holding time. Obtaining these quantities requires solving the boundary value problems (2.3.11) and (2.3.13). Thus, it is nontrivial and beyond the scope of this paper, to generalize Q_e to many dimensions and we stress our main purpose in constructing Q_e is to benchmark other (more practical) approximations in one dimension.

2.4. Realizable Discretizations in 2D

Here we derive realizable finite difference schemes for planar SDEs. The main issue in this part is the discretization of the mixed partial derivatives appearing in (2.1.1), subject to the constraint that the realizability condition (Q2) holds. In fact, a direct discretization of L may lead to an approximation that violates (Q2).

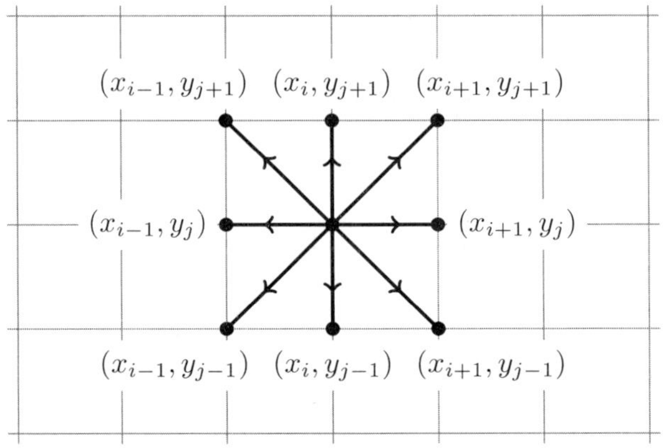

FIGURE 2.4. **Grid in two dimensions.** This diagram illustrates the reaction channels from the point (x_i, y_j) on a gridded space in two dimensions. These channels are used to obtain a realizable discretization of the generator L of the SDE. Note that the channels in the diagonal and antidiagonal directions are needed to approximate mixed partial derivatives that may appear in the continuous generator L. The resulting realizable discretizations can be found in (2.4.6) and (2.4.7).

We use upwinded discretizations of the mixed partial derivatives to get around this issue.

To be specific, consider a planar SDE whose domain Ω is in $\mathbb{R}^2$. Let $M(x, y) = \sigma(x, y)\sigma(x, y)^T$ denote the diffusion matrix for all $(x, y) \in \Omega$, and write the drift and diffusion of this SDE in component form as:

$$\mu(x, y) = \begin{bmatrix} \mu^1(x, y) \\ \mu^2(x, y) \end{bmatrix} \quad \text{and} \quad M(x, y) = \begin{bmatrix} M^{11}(x, y) & M^{12}(x, y) \\ M^{12}(x, y) & M^{22}(x, y) \end{bmatrix} .$$

Let $S = \{(x_i, y_j)\}$ be a collection of grid points on a rectangular grid in two-space as illustrated in Figure 2.4. Given $f : \mathbb{R}^2 \to \mathbb{R}$, we adopt the shorthand notation:

shorthand	meaning
$f_{i,j} = f(x_i, y_j)$	grid point function evaluation
$f_{i\pm1,j} = f(x_{i\pm1}, y_j)$	nearest neighbor grid
$f_{i,j\pm1} = f(x_i, y_{j\pm1})$	point function evaluation
$f_{i\pm1,j\pm1} = f(x_{i\pm1}, y_{j\pm1})$	next to nearest neighbor
$f_{i\pm1,j\mp1} = f(x_{i\pm1}, y_{j\mp1})$	grid point function evaluation

The distance between neighboring vertical or horizontal grid points is permitted to be variable. Hence, it helps to define:

$$\begin{cases} \delta x_i^+ = x_{i+1} - x_i \,, \quad \delta x_i^- = x_i - x_{i-1} \,, \quad \delta x_i = \dfrac{\delta x_i^+ + \delta x_i^-}{2} \,, \\ \delta y_j^+ = y_{j+1} - y_j \,, \quad \delta y_j^- = y_j - y_{j-1} \,, \quad \delta y_j = \dfrac{\delta y_j^+ + \delta y_j^-}{2} \end{cases}$$

at all grid points $(x_i, y_j) \in S$.

We first construct a generalization of the first-order scheme (2.3.7). The drift term in L is approximated using an upwinded finite difference:

$$\mu^1(x_i, y_j)\frac{\partial f}{\partial x}(x_i, y_j) \approx \frac{(\mu^1_{i,j} \vee 0)}{\delta x_i^+}(f_{i+1,j} - f_{i,j}) + \frac{(\mu^1_{i,j} \wedge 0)}{\delta x_i^-}(f_{i,j} - f_{i-1,j}) \tag{2.4.1}$$

and similarly,

$$\mu^2(x_i, y_j)\frac{\partial f}{\partial y}(x_i, y_j) \approx \frac{(\mu^2_{i,j} \vee 0)}{\delta y_j^+}(f_{i,j+1} - f_{i,j}) + \frac{(\mu^2_{i,j} \wedge 0)}{\delta y_j^-}(f_{i,j} - f_{i,j-1}). \tag{2.4.2}$$

This part of the discretization of L clearly satisfies (Q2) and (Q1) with $p = 1$. To discretize the second-order partial derivatives in L, we recruit all of the channels that appear in Figure 2.4 in order. Specifically,

$$\begin{aligned} \operatorname{trace}(M(x_i, y_j)D^2 f(x_i, y_j)) \approx A^{\pm} &\underbrace{(f_{i\pm1,j} - f_{i,j})}_{\text{horizontal differences}} + B^{\pm} \underbrace{(f_{i,j\pm1} - f_{i,j})}_{\text{vertical differences}} \\ &+ C^{\pm}\underbrace{(f_{i\pm1,j\pm1} - f_{i,j})}_{\text{diagonal differences}} + D^{\pm} \underbrace{(f_{i\pm1,j\mp1} - f_{i,j})}_{\text{antidiagonal differences}} \end{aligned} \tag{2.4.3}$$

where we introduced the following labels for the jump rates: $A^{\pm}$, $B^{\pm}$, $C^{\pm}$ and $D^{\pm}$. To be clear there are eight jump rates and eight finite differences in this approximation. We determine the eight rate functions by expanding the horizontal/vertical finite differences in this approximation to obtain

$$\begin{aligned} f_{i\pm1,j} - f_{i,j} &\approx \pm\left(\frac{\partial f}{\partial x}\right)_{i,j} \delta x_i^{\pm} + \frac{1}{2}\left(\frac{\partial^2 f}{\partial x^2}\right)_{i,j} (\delta x_i^{\pm})^2 \\ f_{i,j\pm1} - f_{i,j} &\approx \pm\left(\frac{\partial f}{\partial y}\right)_{i,j} \delta y_j^{\pm} + \frac{1}{2}\left(\frac{\partial^2 f}{\partial y^2}\right)_{i,j} (\delta y_j^{\pm})^2 \end{aligned} \tag{2.4.4}$$

Similarly, we expand the diagonal/antidiagonal differences to obtain

$$\begin{aligned} f_{i\pm1,j\pm1} - f_{i,j} &\approx \pm\left(\frac{\partial f}{\partial x}\right)_{i,j} \delta x_i^{\pm} \pm \left(\frac{\partial f}{\partial y}\right)_{i,j} \delta y_j^{\pm} \\ &+ \frac{1}{2}\left(\frac{\partial^2 f}{\partial x^2}\right)_{i,j} (\delta x_i^{\pm})^2 + \frac{1}{2}\left(\frac{\partial^2 f}{\partial y^2}\right)_{i,j} (\delta y_j^{\pm})^2 \\ &+ \left(\frac{\partial^2 f}{\partial x \partial y}\right)_{i,j} \delta x_i^{\pm} \delta y_j^{\pm} \\ f_{i\pm1,j\mp1} - f_{i,j} &\approx \pm\left(\frac{\partial f}{\partial x}\right)_{i,j} \delta x_i^{\pm} \mp \left(\frac{\partial f}{\partial y}\right)_{i,j} \delta y_j^{\mp} \\ &+ \frac{1}{2}\left(\frac{\partial^2 f}{\partial x^2}\right)_{i,j} (\delta x_i^{\pm})^2 + \frac{1}{2}\left(\frac{\partial^2 f}{\partial y^2}\right)_{i,j} (\delta y_j^{\mp})^2 \\ &- \left(\frac{\partial^2 f}{\partial x \partial y}\right)_{i,j} \delta x_i^{\pm} \delta y_j^{\mp} \end{aligned} \tag{2.4.5}$$

Note that the first order partial derivatives appearing in these expansions appear in plus and minus pairs, and ultimately cancel out. This cancellation is necessary to obtain an approximation to the second-order pure and mixed partial derivatives in (2.4.3). Using these expansions it is straightforward to show that the following choice of jump rates

$$\begin{aligned} D^{\pm} &= -(M_{i,j}^{12} \wedge 0)\,(\delta x_i^{\pm} \delta y_j^{\mp})^{-1} \\ C^{\pm} &= (M_{i,j}^{12} \vee 0)\,(\delta x_i^{\pm} \delta y_j^{\pm})^{-1} \\ B^{\pm} &= M_{i,j}^{22} (\delta y_j^{\pm} \delta y_j)^{-1} - C^{\pm} - D^{\mp} \\ A^{\pm} &= M_{i,j}^{11} (\delta x_i^{\pm} \delta x_i)^{-1} - C^{\pm} - D^{\pm} \end{aligned}$$

imply that the approximation in (2.4.3) is valid. Note here that we have used an upwinded finite difference to approximate the mixed partial derivatives of L, which ensures that the jump rates $C^{\pm}$ and $D^{\pm}$ are always non-negative. However, the price for this upwinded treatment is that we must subtract these jump rates from $A^{\pm}$ and $B^{\pm}$ – more on this point below.

Combining the above approximations to the drift and diffusion terms yields:

$$\begin{aligned} \tilde{Q}_u f_{i,j} =& \left(\frac{\mu_{i,j}^1 \vee 0}{\delta x_i^+} + \frac{M_{i,j}^{11}}{\delta x_i \delta x_i^+} - \frac{M_{i,j}^{12} \vee 0}{\delta x_i^+ \delta y_j^+} + \frac{M_{i,j}^{12} \wedge 0}{\delta x_i^+ \delta y_j^-} \right) (f_{i+1,j} - f_{i,j}) \\ &+ \left(-\frac{\mu_{i,j}^1 \wedge 0}{\delta x_i^-} + \frac{M_{i,j}^{11}}{\delta x_i \delta x_i^-} - \frac{M_{i,j}^{12} \vee 0}{\delta x_i^- \delta y_j^-} + \frac{M_{i,j}^{12} \wedge 0}{\delta x_i^- \delta y_j^+} \right) (f_{i-1,j} - f_{i,j}) \\ &+ \left(\frac{\mu_{i,j}^2 \vee 0}{\delta y_j^+} + \frac{M_{i,j}^{22}}{\delta y_j \delta y_j^+} - \frac{M_{i,j}^{12} \vee 0}{\delta x_i^+ \delta y_j^+} + \frac{M_{i,j}^{12} \wedge 0}{\delta x_i^- \delta y_j^+} \right) (f_{i,j+1} - f_{i,j}) \\ &+ \left(-\frac{\mu_{i,j}^2 \wedge 0}{\delta y_j^-} + \frac{M_{i,j}^{22}}{\delta y_j \delta y_j^-} - \frac{M_{i,j}^{12} \vee 0}{\delta x_i^- \delta y_j^-} + \frac{M_{i,j}^{12} \wedge 0}{\delta x_i^+ \delta y_j^-} \right) (f_{i,j-1} - f_{i,j}) \\ &+ \frac{M_{i,j}^{12} \vee 0}{\delta x_i^+ \delta y_j^+} (f_{i+1,j+1} - f_{i,j}) + \frac{M_{i,j}^{12} \vee 0}{\delta x_i^- \delta y_j^-} (f_{i-1,j-1} - f_{i,j}) \\ &- \frac{M_{i,j}^{12} \wedge 0}{\delta x_i^+ \delta y_j^-} (f_{i+1,j-1} - f_{i,j}) - \frac{M_{i,j}^{12} \wedge 0}{\delta x_i^- \delta y_j^+} (f_{i-1,j+1} - f_{i,j}). \end{aligned} \tag{2.4.6}$$

This infinitesimal generator is a two-dimensional generalization of (2.3.6). However, unlike the one dimensional case, finite differences in the diagonal and antidiagonal directions are used to approximate mixed partial derivatives. The spurious pure derivatives produced by these differences are then subtracted off the jump rates associated to the vertical and horizontal finite differences. This subtraction may affect the realizability condition (Q2), and in particular, the weights in the horizontal and vertical finite differences may become negative. Note that this issue becomes a bit moot if the diffusion matrix is uniformly diagonally dominant, since this property implies that

$$M_{i,j}^{11} \wedge M_{i,j}^{22} \ge |M_{i,j}^{12}| \qquad \text{for all } (x_i, y_j) \in S$$

and hence, $\tilde{Q}_u$ satisfies the realizability condition (Q2) on a gridded state space, with evenly spaced points in the x and y directions. In general, however, a 2×2 symmetric positive definite matrix (like $M(x)$) is weakly diagonally dominant (see Definition 2.8.1), and therefore, varying the spatial step sizes in $\tilde{Q}_u$ may be necessary to ensure that (Q2) holds. In fact, we will show that this is always possible to do for planar SDEs.

Next we construct a generalization of the second-order scheme (2.3.10). To do this, we approximate the first and second-order 'pure' partial derivatives in L using a weighted central difference, and use an upwinded finite difference to treat the mixed partials, we obtain the following generator:
(2.4.7)

$$
\begin{aligned}
\tilde{Q}_c f_{i,j} &= \exp\left(+\frac{1}{2}\tilde{\mu}^1_{i,j}\delta x_i^+\right)\left(\frac{M^{11}_{i,j}}{\delta x_i \delta x_i^+} - \frac{M^{12}_{i,j}\vee 0}{\delta x_i^+ \delta y_j^+} + \frac{M^{12}_{i,j}\wedge 0}{\delta x_i^+ \delta y_j^-}\right)(f_{i+1,j} - f_{i,j}) \\
&+ \exp\left(-\frac{1}{2}\tilde{\mu}^1_{i,j}\delta x_i^-\right)\left(\frac{M^{11}_{i,j}}{\delta x_i \delta x_i^-} - \frac{M^{12}_{i,j}\vee 0}{\delta x_i^- \delta y_j^-} + \frac{M^{12}_{i,j}\wedge 0}{\delta x_i^- \delta y_j^+}\right)(f_{i-1,j} - f_{i,j}) \\
&+ \exp\left(+\frac{1}{2}\tilde{\mu}^2_{i,j}\delta y_j^+\right)\left(\frac{M^{22}_{i,j}}{\delta y_j \delta y_j^+} - \frac{M^{12}_{i,j}\vee 0}{\delta x_i^+ \delta y_j^+} + \frac{M^{12}_{i,j}\wedge 0}{\delta x_i^- \delta y_j^+}\right)(f_{i,j+1} - f_{i,j}) \\
&+ \exp\left(-\frac{1}{2}\tilde{\mu}^2_{i,j}\delta y_j^-\right)\left(\frac{M^{22}_{i,j}}{\delta y_j \delta y_j^-} - \frac{M^{12}_{i,j}\vee 0}{\delta x_i^- \delta y_j^-} + \frac{M^{12}_{i,j}\wedge 0}{\delta x_i^+ \delta y_j^-}\right)(f_{i,j-1} - f_{i,j}) \\
&+ \exp\left(+\frac{1}{2}\tilde{\mu}^1_{i,j}\delta x_i^+ + \frac{1}{2}\tilde{\mu}^2_{i,j}\delta y_j^+\right)\frac{M^{12}_{i,j}\vee 0}{\delta x_i^+ \delta y_j^+}(f_{i+1,j+1} - f_{i,j}) \\
&+ \exp\left(-\frac{1}{2}\tilde{\mu}^1_{i,j}\delta x_i^- - \frac{1}{2}\tilde{\mu}^2_{i,j}\delta y_j^-\right)\frac{M^{12}_{i,j}\vee 0}{\delta x_i^- \delta y_j^-}(f_{i-1,j-1} - f_{i,j}) \\
&- \exp\left(+\frac{1}{2}\tilde{\mu}^1_{i,j}\delta x_i^+ - \frac{1}{2}\tilde{\mu}^2_{i,j}\delta y_j^-\right)\frac{M^{12}_{i,j}\wedge 0}{\delta x_i^+ \delta y_j^-}(f_{i+1,j-1} - f_{i,j}) \\
&- \exp\left(-\frac{1}{2}\tilde{\mu}^1_{i,j}\delta x_i^- + \frac{1}{2}\tilde{\mu}^2_{i,j}\delta y_j^+\right)\frac{M^{12}_{i,j}\wedge 0}{\delta x_i^- \delta y_j^+}(f_{i-1,j+1} - f_{i,j}).
\end{aligned}
$$

Here we have introduced the transformed drift field $\tilde{\mu} = (\tilde{\mu}^1, \tilde{\mu}^2)^T$, which is defined pointwise as the solution to the linear system

$$M_{i,j}\tilde{\mu}_{i,j} = \mu_{i,j} \qquad \text{for all } (x_i, y_j) \in S.$$

Like $\tilde{Q}_u$, if the diffusion matrix is diagonally dominant, then this generator is realizable on an evenly spaced grid. In §3.9 and §3.10, we test $\tilde{Q}_c$ on planar SDE problems. In these problems the diffusion matrices are not diagonally dominant, and we must use variable step sizes in order to obtain a realizable discretization. Corollary 2.8.3 states that in two-dimensional problems, one can always find variable spatial step sizes in $\tilde{Q}_c$ so that (Q2) holds. In §2.8, we generalize this result to multiple dimensions and prove that if the diffusion matrix is weakly diagonally dominant, then one can always find variable spatial step sizes such that this generalized approximation of L is realizable.

To avoid assuming the diffusion matrix is diagonally dominant, which is too restrictive, we next consider approximations whose state space is possibly gridless; refer back to §2.2 for our definition of a gridless state space.

2.5. Realizable Discretizations in nD

We now present a realizable discretization on a gridless state space contained in $\Omega \subset \mathbb{R}^n$. To introduce this approximation, let $\{e_i\}_{i=1}^n$ be the standard basis of $\mathbb{R}^n$, set the number of reaction channel (or jump) directions equal to n, and let i

be an index over these reaction channel directions $1 \le i \le n$. We will allow forward/backward jumps along each channel direction, and since there are n reaction channel directions, the total number of reaction channels (or jump states) is $2n$. Let $h_i^{\pm}(x)$ be forward/backward spatial step sizes at state x, let $h_i(x)$ be the average step size at state x defined as

$$h_i(x) = (h_i^+(x) + h_i^-(x))/2 \ ,$$

and let $\{\sigma_i(x) = \sigma e_i\}_{i=1}^n$ be the columns of the noise matrix $\sigma(x)$. We set the ith channel direction to be $\sigma_i(x)$ for $1 \le i \le n$. Finally, let $\tilde{\mu}(x)$ be a transformed drift field defined pointwise as:

$$M(x)\tilde{\mu}(x) = \mu(x) \tag{2.5.1}$$

where $M(x) = \sigma(x)\sigma(x)^T$ is the $n \times n$ diffusion matrix. If $M(x)$ is positive definite, then this linear system has a unique solution $\tilde{\mu}(x)$ for every $x \in \mathbb{R}^n$. This condition holds in the interior of Ω for all of the SDE problems we consider in this paper. With this notation in hand, here is a second order accurate, realizable discretization in n dimensions.

(2.5.2)

$$\boxed{\begin{aligned} Q_c f(x) = \sum_{i=1}^n & \frac{1}{h_i^+(x) h_i(x)} \exp\left(\frac{h_i^+(x)}{2} \tilde{\mu}(x)^T \sigma_i(x) \right) \left(f\left(x + h_i^+(x)\sigma_i(x)\right) - f(x) \right) \\ & + \frac{1}{h_i^-(x) h_i(x)} \exp\left(-\frac{h_i^-(x)}{2} \tilde{\mu}(x)^T \sigma_i(x) \right) \left(f\left(x - h_i^-(x)\sigma_i(x)\right) - f(x) \right) \end{aligned}}$$

This approximation uses $2n$ weighted, variable step size finite differences along the directions given by the columns of the noise coefficient matrix evaluated at the state x. When the noise coefficient matrix is a positive scalar multiple of the identity matrix, then the state space of this generator is gridded and this generator can be represented by a matrix. In general, we stress that the state space of this approximation is gridless, and thus the generator cannot be represented by a matrix. Nevertheless this approximation of L generates a Markov process, whose sample paths can be produced by using the SSA given in Algorithm 2.1. A realization of this process is a continuous-time random walk with user-defined jump size(s), jump directions given by the columns of the noise coefficient matrix, and a mean holding time that is determined by the discretized generator. This generator extends realizable spatial discretizations appearing in previous work [**40, 102, 120, 121, 160**] to the SDE (2.1.2), which generically has a non-symmetric infinitesimal generator L and multiplicative noise. We stress that the field of spatial step sizes entering (2.5.2) can be chosen such that the jump size in (Step 2) of Algorithm 2.1 is uniformly bounded.

The next proposition states that this approximation satisfies (Q1) and (Q2).

PROPOSITION 2.5.1. *Assuming that:*

$$|h_i^+(x) - h_i^-(x)| \vee (h_i^+(x))^2 \vee (h_i^-(x))^2 \le h^2 \quad \textit{for all } x \in \Omega$$

then the operator Q_c in (2.5.2) satisfies (Q2) and (Q1) with $p = 2$.

PROOF. This proof is somewhat informal. A formal proof requires hypotheses on the SDE coefficients and the test function $f(x)$. For clarity of presentation, we defer these technical issues to Chapter 4. Here we concentrate on showing in simple terms that the approximation meets the requirements (Q1) and (Q2). Note that the jump rates appearing in Q_c are all positive, and thus, the realizability condition (Q2) is satisfied.

To check second-order accuracy or (Q1) with $p = 2$, we require the following notation: for any $k \geq 1$, let $D^k f(x)(u_1, \ldots, u_k)$ denote the kth-derivative of $f(x)$ applied to the vectors $u_1, \ldots, u_k \in \mathbb{R}^n$. A Maclaurin expansion of the forward/backward finite differences appearing in the function $Q_c f(x)$ yields:

$$\begin{aligned} Q_c f(x) = \sum_{i=1}^{n} \frac{1}{h_i(x)} & Df(x)(\sigma_i(x)) \left(\exp\left(\tilde{\mu}(x)^T \sigma_i(x) \frac{h_i^+(x)}{2} \right) \right. \\ & \left. - \exp\left(-\tilde{\mu}(x)^T \sigma_i(x) \frac{h_i^-(x)}{2} \right) \right) \\ + \frac{1}{2h_i(x)} & D^2 f(x)(\sigma_i(x), \sigma_i(x)) \left(h_i^+(x) \exp\left(\tilde{\mu}(x)^T \sigma_i(x) \frac{h_i^+(x)}{2} \right) \right. \\ & \left. + h_i^-(x) \exp\left(-\tilde{\mu}(x)^T \sigma_i(x) \frac{h_i^-(x)}{2} \right) \right) \\ + \frac{1}{6h_i(x)} & D^3 f(x)(\sigma_i(x), \sigma_i(x), \sigma_i(x)) \left(h_i^+(x)^2 \exp\left(\tilde{\mu}(x)^T \sigma_i(x) \frac{h_i^+(x)}{2} \right) \right. \\ & \left. - h_i^-(x)^2 \exp\left(-\tilde{\mu}(x)^T \sigma_i(x) \frac{h_i^-(x)}{2} \right) \right) + O(h^2). \end{aligned}$$

In order to obtain a crude estimate of the remainder term in this expansion, we assume that f has four bounded derivatives; and we use our hypotheses on the fields of forward/backward spatial step sizes, and the identity $a^3 + b^3 = (a+b)(a^2 + b^2 - ab)$ with $a = h_i^+(x)$ and $b = h_i^-(x)$ combined with the relation $h_i(x) = (h_i^+(x) + h_i^-(x))/2$.

A second Maclaurin expansion of the exponentials appearing above yields,

$$\begin{aligned} Q_c f(x) &= \sum_{i=1}^{n} Df(x)^T \sigma_i(x) \tilde{\mu}(x)^T \sigma_i(x) + D^2 f(x)(\sigma_i(x), \sigma_i(x)) + O(h^2) \\ &= \operatorname{trace} \left(\left(Df(x) \tilde{\mu}(x)^T + D^2 f(x) \right) \underbrace{\sum_{i=1}^{n} \sigma_i \sigma_i^T}_{= M(x)} \right) + O(h^2) \\ &= \operatorname{trace} \left(Df(x) \mu(x)^T + D^2 f(x) M(x) \right) + O(h^2) \\ &= Lf(x) + O(h^2) \end{aligned}$$

where we used the cyclic property of the trace operator, the basic linear algebra fact that the product $\sigma \sigma^T$ is the sum of the outer products of the columns of σ, and elementary identities to estimate the remainder terms, like $a^2 - b^2 = (a+b)(a-b)$ with $a = h_i^+(x)$ and $b = h_i^-(x)$. Thus, Q_c is second-order accurate as claimed. □

The analysis in Chapter 4 focusses on the stability, accuracy and complexity of the generator Q_c, mainly because it is second-order accurate. However, let us briefly consider two simpler alternatives, which are however first-order accurate. To avoid evaluating the exponential function appearing in the transition rates of

Q_c in (2.5.2), an upwinded generator is useful

$$
\begin{aligned}
Q_u f(x) = \sum_{i=1}^{n} & \frac{(\tilde{\mu}(x)^T \sigma_i(x)) \vee 0}{h_i^+(x)} \left(f\left(x + h_i^+(x)\sigma_i(x)\right) - f(x)\right) \\
& - \frac{(\tilde{\mu}(x)^T \sigma_i(x)) \wedge 0}{h_i^-(x)} \left(f\left(x - h_i^-(x)\sigma_i(x)\right) - f(x)\right) \\
& + \frac{1}{h_i^+(x) h_i(x)} \left(f\left(x + h_i^+(x)\sigma_i(x)\right) - f(x)\right) \\
& + \frac{1}{h_i^-(x) h_i(x)} \left(f\left(x - h_i^-(x)\sigma_i(x)\right) - f(x)\right) .
\end{aligned} \tag{2.5.3}
$$

To avoid solving the linear system in (2.5.1), an upwinded generator that uses the standard basis to discretize the drift term in L is useful

$$
\begin{aligned}
Q_{uu} f(x) = \sum_{i=1}^{n} & \frac{(\mu(x)^T e_i) \vee 0}{h_i^+(x)} \left(f\left(x + h_i^+(x)e_i\right) - f(x)\right) \\
& - \frac{(\mu(x)^T e_i) \wedge 0}{h_i^-(x)} \left(f\left(x - h_i^-(x)e_i\right) - f(x)\right) \\
& + \frac{1}{h_i^+(x) h_i(x)} \left(f\left(x + h_i^+(x)\sigma_i(x)\right) - f(x)\right) \\
& + \frac{1}{h_i^-(x) h_i(x)} \left(f\left(x - h_i^-(x)\sigma_i(x)\right) - f(x)\right) .
\end{aligned} \tag{2.5.4}
$$

The following proposition states that these approximations define first-order accurate, infinitesimal generators.

PROPOSITION 2.5.2. *Assuming that:*

$$|h_i^+(x) - h_i^-(x)| \vee h_i^+(x) \vee h_i^-(x) \le h \quad \textit{for all } x \in \Omega$$

then the operators Q_u and Q_{uu} satisfy (Q2) and (Q1) with $p = 1$.

The proof of Proposition 2.5.2 is similar to the proof of Proposition 2.5.1 and therefore omitted.

2.6. Scaling of Approximation with System Size

Here we discuss the cost of the approximation given in (2.5.2). In §4.3, we quantify the cost of producing sample paths using the SSA method as a function of the spatial step size parameter, i.e., as $h \to 0$. In this section, we focus on the cost of the approximation as $n \to \infty$.

First, to provide a baseline for comparison, let us briefly review the typical cost of a standard numerical PDE solver for the Kolmogorov equation. To setup this method, the domain of the underlying SDE is typically discretized using a grid consisting of a large number of grid points. If the domain is bounded, then this number of grid points is typically exponential in n. A discretization matrix is then constructed over this grid, where each row of this matrix typically corresponds to an approximation of L evaluated at a grid point. Even though the discretization matrix associated to standard PDE methods (like finite difference, volume and element methods) is often sparse, the computational effort required to build the entries of this matrix and compute matrix-vector products in order to evolve the resulting semi-discrete ODE (either exactly or approximately) scales exponentially with n. If the domain is unbounded, the situation is worse: this ODE is infinite-dimensional and a standard numerical PDE method does not directly apply.

In contrast, by using Monte-Carlo to construct the approximation, we can compute a solution to the Kolmogorov equation at an arbitrary point in space with a cost that is much lower than exponential in n. Indeed, note from Algorithm 2.1 that an SSA simulation of the Markov process induced by the infinitesimal generator in (2.5.2) takes as input the reaction rates and channels, which can be peeled off of (2.5.2). More to the point, both the spatial and temporal updates in each SSA step use only local information. Put another way, in the case of a gridded state space, a step of the SSA method only requires computing a row of a typically sparse discretization matrix associated to the infinitesimal generator, not the entire matrix. Thus, the cost of each SSA step in high dimension is much less severe than standard numerical PDE solvers and remains practical. We stress that this Monte-Carlo method is used to construct a local solution to the Kolmogorov equation.

Next consider the average number of SSA steps required to reach a finite time, and how this quantity depends on n. Referring to Algorithm 2.1, the random time increment δt gives the amount of time the process spends in state $X(t)$ before jumping to a new state $X(t+\delta t)$. In each step of the SSA method, only a single degree of freedom is modified because the process moves from the current position to a new position in the direction of only one of the n columns of the noise matrix. In other words, the approximation evolves by local moves, and in this way, it is related to splitting methods for SDEs and ODEs based on local moves, for example the per-particle splitting integrator proposed for Metropolis-corrected Langevin integrators in [**19**] or J-splitting schemes for Hamiltonian systems [**42**]. However, the physical clock in splitting algorithms does not tick with each local move as it does in (Step 1) of the SSA method. Instead the physical time in a splitting scheme elapses only after all of the local moves have happened. In contrast, the physical clock ticks with each step of the SSA method, and we stress that only a single degree of freedom is modified per step. Thus, accuracy typically requires that the mean holding time is inversely proportional to n.

To illustrate this point, consider approximating n decoupled SDEs using the SSA induced by (2.5.2). In this situation, the mean time step for a global move of the system – where each component has taken a step – cannot depend on dimension because the underlying SDEs are decoupled. For simplicity, assume the underlying SDE for the ith approximation is $dY_i = \mu(Y_i)dt + \sqrt{2}dW_i$ for $1 \le i \le n$, where $\{W_i\}_{i=1}^n$ are n independent Brownian motions; and assume that the forward/backward spatial step sizes in every reaction channel direction are uniform and equal to h. Write a vector x in components as $x = (x_1, \dots, x_n)^T$. Then, from (2.5.2) the total jump rate from state x can be written as:

$$\lambda(x) = \sum_{i=1}^{n} \frac{2}{h^2} \cosh\left(\mu(x_i)\frac{h}{2}\right).$$

Expanding $\lambda(x)$ about $h = 0$ and organizing this expansion in powers of h yields

$$\lambda(x) = \frac{2n}{h^2} + \frac{1}{4}\sum_{i=1}^{n} \mu(x_i)^2 + O(h^2).$$

Similarly, expanding the mean holding time about $h = 0$ yields

$$\lambda(x)^{-1} = \frac{h^2}{2n} - \frac{h^4}{16n^2} \sum_{i=1}^{n} \mu(x_i)^2 + O(h^6).$$

Thus, to leading order we see that the mean holding time scales like $O(h^2/(2n))$, as expected.

To summarize, if there are n linearly independent, jump directions, then each step of the SSA involves an update in only a single degree of freedom, and hence, accuracy necessitates that the time elapsed between jumps is inversely proportional with n. Thus, the overall complexity of the SSA method is no different from a splitting temporal integrator for this problem that uses local moves. While the precise cost of this approximation is problem-specific and may vary strongly with changing drift and noise coefficients, the scaling of this approximation is typically much less than exponential.

2.7. Generalization of Realizable Discretizations in nD

In §2.8, we will construct realizable discretizations on gridded state spaces when the diffusion matrix of the SDE is weakly diagonally dominant. As a preliminary step in this direction, in this section we introduce a more general decomposition of the diffusion matrix and we use this decomposition to extend the discretization given in §2.5.

Let s be a natural number. Suppose the diffusion matrix has the following rank-one expansion:

$$M(x) = \sum_{i=1}^{s} w_i(x) \eta_i(x) \eta_i(x)^T \tag{2.7.1}$$

where $\eta_i(x) \in \mathbb{R}^n$ is an n-vector and $w_i(x) \geq 0$ is a non-negative weight function for all $x \in \Omega$, and for $1 \leq i \leq s$. Since the diffusion matrix $M(x)$ is assumed to be globally positive definite, it must be that $s \geq n$ and that the set of vectors $\{\eta_i(x)\}_{i=1}^{s}$ spans $\mathbb{R}^n$ for every $x \in \mathbb{R}^n$. As a simple example, since $M(x) = \sigma(x)\sigma(x)^T$, the diffusion matrix can always be put in the form of (2.7.1) with $s = n$, unit weights $w_i(x) = 1$, and $\eta_i(x) = \sigma_i(x)$ for $1 \leq i \leq n$, or equivalently,

$$M(x) = \sum_{i=1}^{n} \sigma_i(x) \sigma_i(x)^T.$$

The following generator generalizes (2.5.2) to diffusion matrices of the form (2.7.1).

(2.7.2)

$$\boxed{\begin{aligned} \mathcal{Q}_c f(x) = \sum_{i=1}^{s} & \frac{w_i(x)}{h_i^+(x) h_i(x)} \exp\left(\frac{h_i^+(x)}{2} \tilde{\mu}(x)^T \eta_i(x) \right) \left(f\left(x + h_i^+(x)\eta_i(x)\right) - f(x) \right) \\ & + \frac{w_i(x)}{h_i^-(x) h_i(x)} \exp\left(-\frac{h_i^-(x)}{2} \tilde{\mu}(x)^T \eta_i(x) \right) \left(f\left(x - h_i^-(x)\eta_i(x)\right) - f(x) \right). \end{aligned}}$$

Recall that, $h_i^{\pm}(x)$ represent the forward/backward spatial step sizes along the ith channel direction; and, $\tilde{\mu}$ is the transformed drift field introduced in (2.5.1). This approximation uses $2n$ weighted, variable step size finite differences along the directions given by the vectors: $\{\eta_i(x)\}_{i=1}^s$. The next proposition states that this generalization of Q_c meets the requirements (Q1) and (Q2).

PROPOSITION 2.7.1. *Suppose that the diffusion matrix satisfies* (2.7.1). *Assuming that:*

$$|h_i^+(x) - h_i^-(x)| \vee (h_i^+(x))^2 \vee (h_i^-(x))^2 \le h^2 \quad \textit{for all } x \in \Omega$$

then the operator $\mathcal{Q}_c$ *in* (2.7.2) *satisfies (Q2) and (Q1) with* $p = 2$.

PROOF. The operator $\mathcal{Q}_c$ satisfies the realizability condition (Q2) since $w_i(x) \ge 0$ by hypothesis on the diffusion field $M(x)$. To check accuracy, we expand in a Maclaurin series the forward/backward differences appearing in $\mathcal{Q}_c f(x)$ to obtain:

$$\mathcal{Q}_c f(x) = \sum_{i=1}^s \frac{w_i}{h_i} Df^T \eta_i \left(\exp\left(\tilde{\mu}^T \eta_i \frac{h_i^+}{2} \right) - \exp\left(-\tilde{\mu}^T \eta_i \frac{h_i^-}{2} \right) \right)$$
$$+ \frac{w_i}{2h_i} D^2 f(\eta_i, \eta_i) \left(h_i^+ \exp\left(\tilde{\mu}^T \eta_i \frac{h_i^+}{2} \right) + h_i^- \exp\left(-\tilde{\mu}^T \eta_i \frac{h_i^-}{2} \right) \right)$$
$$+ \frac{w_i}{6h_i} D^3 f(\eta_i, \eta_i, \eta_i) \left((h_i^+)^2 \exp\left(\tilde{\mu}^T \eta_i \frac{h_i^+}{2} \right) - (h_i^-)^2 \exp\left(-\tilde{\mu}^T \eta_i \frac{h_i^-}{2} \right) \right) + O(h^2).$$

A similar Maclaurin expansion of the exponentials yields,

$$\begin{aligned}
\mathcal{Q}_c f(x) &= \sum_{i=1}^s w_i Df^T \eta_i \tilde{\mu}^T \eta_i + w_i D^2 f(\eta_i, \eta_i) + O(h^2) \\
&= \operatorname{trace} \left((Df \tilde{\mu}^T + D^2 f) \left(\underbrace{\sum_{i=1}^s w_i \eta_i \eta_i^T}_{=M(x)} \right) \right) + O(h^2) \\
&= \operatorname{trace} \left(Df(x)\, \mu(x)^T + D^2 f(x)\, M(x) \right) + O(h^2) \\
&= Lf(x) + O(h^2)
\end{aligned}$$

where we used (2.7.1). Thus, $\mathcal{Q}_c$ is second-order accurate as claimed. □

As an application of Proposition 2.7.1 we next show that realizable discretizations on gridded state spaces are possible to construct if the diffusion matrix of the SDE is diagonally dominant.

2.8. Weakly Diagonally Dominant Case

Here we extend the approximation $\tilde{Q}_c$ in (2.4.7) from two to many dimensional SDEs. We show that if the diffusion matrix is weakly diagonally dominant, and under a suitable choice of spatial step sizes, this generalization is both second-order accurate and realizable. Before we present our generalization, let us quickly recall what it means for a matrix to be weakly diagonally dominant.

DEFINITION 2.8.1. An $n \times n$ matrix A is *weakly diagonally dominant* if there exists an n-vector $y = (y_1, \ldots, y_n)^T$ with positive entries such that

$$|A_{ii}| y_i \ge \sum_{j \ne i} |A_{ij}| y_j \tag{2.8.1}$$

for all $1 \le i \le n$.

Note that diagonal dominance of A requires that y_i in (2.8.1) equals unity for all $1 \le i \le n$. The following lemma is straightforward to derive from this definition, and will be useful in the sequel.

LEMMA 2.8.1. *Let A be weakly diagonally dominant. Then there exists a positive diagonal matrix P such that the square matrix $\tilde{A} = PAP$ is diagonally dominant.*

Let $\Delta(x) = (\Delta_1(x), \ldots, \Delta_n(x))^T$ be a field of spatial step sizes, which permit a different spatial step size for each coordinate axis. Let $\tilde{\mu}$ be the transformed drift field defined in (2.5.1). Here is a multi-dimensional generalization of $\tilde{Q}_c$

$$\tilde{Q}_c f(x) = \lambda(x)\, \mathbb{E}(f(\xi_x) - f(x)) \tag{2.8.2}$$

where the expectation is taken over the random vector ξ_x which has a discrete probability distribution given by:

$$\begin{aligned}
\xi_x \sim \lambda(x)^{-1} \Bigg(& \sum_{i=1}^{n} \delta(y_i^+) \exp\left(\frac{1}{2}\tilde{\mu}^T (y_i^+ - x)\right) \left(\frac{M_{ii}}{\Delta_i^2} - \sum_{j \ne i} \frac{|M_{ij}|}{\Delta_i \Delta_j}\right) \\
& + \sum_{i=1}^{n} \delta(y_i^-) \exp\left(\frac{1}{2}\tilde{\mu}^T (y_i^- - x)\right) \left(\frac{M_{ii}}{\Delta_i^2} - \sum_{j \ne i} \frac{|M_{ij}|}{\Delta_i \Delta_j}\right) \\
& + \sum_{\substack{1 \le i \le n \\ j > i}} \delta(y_{ij}^{+,d}) \exp\left(\frac{1}{2}\tilde{\mu}^T (y_{ij}^{+,d} - x)\right) \left(\frac{M_{ij} \vee 0}{\Delta_i \Delta_j}\right) \\
& + \sum_{\substack{1 \le i \le n \\ j > i}} \delta(y_{ij}^{-,d}) \exp\left(\frac{1}{2}\tilde{\mu}^T (y_{ij}^{-,d} - x)\right) \left(\frac{M_{ij} \vee 0}{\Delta_i \Delta_j}\right) \\
& - \sum_{\substack{1 \le i \le n \\ j > i}} \delta(y_{ij}^{+,a}) \exp\left(\frac{1}{2}\tilde{\mu}^T (y_{ij}^{+,a} - x)\right) \left(\frac{M_{ij} \wedge 0}{\Delta_i \Delta_j}\right) \\
& - \sum_{\substack{1 \le i \le n \\ j > i}} \delta(y_{ij}^{-,a}) \exp\left(\frac{1}{2}\tilde{\mu}^T (y_{ij}^{-,a} - x)\right) \left(\frac{M_{ij} \wedge 0}{\Delta_i \Delta_j}\right) \Bigg)
\end{aligned} \tag{2.8.3}$$

where $\delta(y)$ is a point mass concentrated at y, $\lambda(x)$ is a normalization constant ($\lambda(x)^{-1}$ is the mean holding time of the approximation in state x), and we have introduced the following reaction channels (or jump states):

$$\begin{cases}
y_i^{\pm}(x) = x \pm e_i \Delta_i\,, & 1 \le i \le n \\
y_{ij}^{\pm,d}(x) = x \pm e_i \Delta_i \pm e_j \Delta_j\,, & 1 \le i \le n\,,\ j > i \\
y_{ij}^{\pm,a}(x) = x \pm e_i \Delta_i \mp e_j \Delta_j\,, & 1 \le i \le n\,,\ j > i
\end{cases} \tag{2.8.4}$$

The superscripts d and a in the reaction channels above refer to the diagonal and anti-diagonal directions in the ij-plane, respectively. For the sake of readability, we suppressed the dependence on x in the functions appearing on the right-hand side of (2.8.3).

The following proposition gives a sufficient condition such that $\tilde{Q}_c$ satisfies (Q1) and (Q2).

PROPOSITION 2.8.2. *Consider the generator $\tilde{Q}_c$ given in (2.8.2). If for all $x \in \mathbb{R}^n$, the $n \times n$ diffusion matrix $M(x)$ is weakly diagonally dominant, then there exists a field Δ such that $\tilde{Q}_c$ in (2.8.2) is realizable and second-order accurate.*

PROOF. This proof is an application of Proposition 2.7.1. By Lemma 2.8.1, there exists a positive diagonal matrix $P(x)$ such that:

$$\tilde{M}(x) = P(x)M(x)P(x) \tag{2.8.5}$$

is diagonally dominant for all $x \in \mathbb{R}^n$. Let $P_{ii}(x)$ denote the ith diagonal entry of the matrix $P(x)$, and define a field of spatial step sizes Δ pointwise by setting $\Delta_i(x) = h/P_{ii}(x)$ for $1 \le i \le n$ and for all $x \in \mathbb{R}^n$, where h is a spatial step size parameter. Hence, we have that:

$$\begin{aligned} M &= \sum_{i=1}^{n} \left(M_{ii} e_i e_i^T + \sum_{j>i} M_{ij}(e_i e_j^T + e_j e_i^T) \right) \\ &= \sum_{i=1}^{n} (\tilde{M}_{ii} - \sum_{j \ne i} |\tilde{M}_{ij}|) \frac{(y_i^+ - x)}{h} \frac{(y_i^+ - x)}{h}^T \\ &+ \sum_{\substack{1 \le i \le n \\ j>i}} (\tilde{M}_{ij} \vee 0) \, \frac{(y_{ij}^{+,d} - x)}{h} \frac{(y_{ij}^{+,d} - x)}{h}^T \\ &- \sum_{\substack{1 \le i \le n \\ j>i}} (\tilde{M}_{ij} \wedge 0) \, \frac{(y_{ij}^{+,a} - x)}{h} \frac{(y_{ij}^{+,a} - x)}{h}^T \end{aligned}$$

where we used the forward reaction channels defined in (2.8.4). Note that the coefficient of each rank one matrix in this last expression is non-negative because $\tilde{M}(x)$ is diagonally dominant. Thus, under the assumption of weak diagonal dominance, we have put $M(x)$ in the form of (2.7.1) with $s = n^2$, and note that $\tilde{Q}_c$ is a special case of Q_c in (2.7.2) with $h_i^{\pm}(x) = h$ and reaction channel directions given by:

$$\left\{ \frac{e_i}{P_{ii}}, \left\{ \frac{e_i}{P_{ii}} + \frac{e_j}{P_{jj}} \right\}, \left\{ \frac{e_i}{P_{ii}} - \frac{e_j}{P_{jj}} \right\} \mid 1 \le i \le n \,, j > i \right\}.$$

The desired result follows from Proposition 2.7.1. □

Weak diagonal dominance of $M(x)$ is not easy to directly verify. However, according to Theorem 9 of [**10**], this property can be determined from a Cholesky factorization of $M(x)$. Indeed, a symmetric positive definite matrix is weakly diagonally dominant if and only if its Cholesky factor has at most two nonzero entries in every column. As an immediate corollary, in two-dimensions every symmetric positive definite matrix is weakly diagonally dominant, and we have the following.

COROLLARY 2.8.3. *Consider the generator $\tilde{Q}_c$ given in (2.8.2) with $n = 2$. Then there exists a field Δ such that $\tilde{Q}_c$ in (2.8.2) is realizable and second-order accurate. Moreover, there exist fields $\delta x^{\pm}$ and $\delta y^{\pm}$ in $\tilde{Q}_c$ in (2.4.7) such that $\tilde{Q}_c$ is realizable and second-order accurate.*

If $M(x)$ is diagonally dominant, then $\tilde{\mathcal{Q}}_c$ satisfies (Q2) on a simple gridded state space, as the next corollary states.

COROLLARY 2.8.4. *If for all* $x \in \mathbb{R}^n$, *the* $n \times n$ *diffusion matrix* $M(x)$ *is diagonally dominant, then* $\tilde{\mathcal{Q}}_c$ *in* (2.8.2) *is realizable on a gridded state space in* $\mathbb{R}^n$ *and second-order accurate.*

As far as we can tell, realizable approximations like $\tilde{\mathcal{Q}}_c$ on a gridded state space were first developed by H. Kushner [**88–98**].

PROOF. In this case the diffusion matrix can be written as:

$$\begin{aligned} M = & \sum_{i=1}^{n} \left(M_{ii} - \sum_{j \neq i} |M_{ij}| \right) e_i e_i^T \\ & + \sum_{\substack{1 \le i \le n \\ j > i}} (M_{ij} \vee 0) \, (e_i + e_j)(e_i + e_j)^T \\ & - \sum_{\substack{1 \le i \le n \\ j > i}} (M_{ij} \wedge 0) \, (e_i - e_j)(e_i - e_j)^T \end{aligned}$$

and the coefficient of each rank one matrix in this last expression is non-negative because $M(x)$ is diagonally dominant by hypothesis. Thus, under the assumption of diagonal dominance, we have put $M(x)$ in the form of (2.7.1) with $s = n^2$, and note that $\tilde{\mathcal{Q}}_c$ is a special case of $\mathcal{Q}_c$ in (2.7.2) with $h_i^{\pm}(x) = h$ and reaction channel directions given by:

$$\left\{ e_i, \left\{ e_i + e_j \right\}, \left\{ e_i - e_j \right\} \bigm| 1 \le i \le n \, , j > i \right\}.$$

Note also that the state space is gridded since the jumps are either to nearest neighbor grid points: $x \pm h e_i$ for all $1 \le i \le n$, or to next to nearest neighbor grid points: $x \pm h e_i \pm h e_j$ and $x \pm h e_i \mp h e_j$ for all $1 \le i \le n$ and $j > i$. $\square$

CHAPTER 3

Examples & Applications

3.1. Introduction

In this chapter we apply realizable discretizations to

(1) a cubic oscillator in 1D with additive noise as described in §3.2;
(2) a log-normal process in 1D with multiplicative noise as described in §3.5;
(3) the Cox-Ingersoll-Ross process in 1D with multiplicative noise as described in §3.6;
(4) a non-symmetric Ornstein-Uhlenbeck process as described in §3.7.1;
(5) the Maier-Stein SDE in 2D with additive noise as described in §3.7.2;
(6) simulation of a particle in a planar square-well potential well in 2D with additive noise as described in §3.7.3;
(7) simulation of a particle in a worm-like chain potential well in 2D with additive noise as described in §3.7.4;
(8) a log-normal process in 2D with multiplicative noise as described in §3.9;
(9) a Lotka-Volterra process in 2D with multiplicative noise as described in §3.10; and,
(10) a Brownian dynamics simulation of the collapse of a colloidal cluster in 39D with and without hydrodynamic interactions as described in §3.11.

For the 1D SDE problems, we numerically test two realizable spatial discretizations with a gridded state space. The first uses an upwinded (resp. central) finite difference method to approximate the first (resp. second) order derivative in (2.1.1).

$$
(3.1.1)\qquad \begin{cases} (\tilde{Q}_u)_{i,i+1} = \dfrac{1}{\delta x_i^+}\left((\mu_i \vee 0) + \dfrac{M_i}{\delta x_i}\right) \\ (\tilde{Q}_u)_{i,i-1} = \dfrac{1}{\delta x_i^-}\left(-(\mu_i \wedge 0) + \dfrac{M_i}{\delta x_i}\right) \end{cases}
$$

The second generator uses a weighted central finite difference scheme to approximate the derivatives in (2.1.1).

$$
(3.1.2)\qquad \begin{cases} (\tilde{Q}_c)_{i,i+1} = \dfrac{1}{\delta x_i\, \delta x_i^+} M_i \exp\left(\dfrac{\mu_i}{M_i}\dfrac{\delta x_i^+}{2}\right) \\ (\tilde{Q}_c)_{i,i-1} = \dfrac{1}{\delta x_i\, \delta x_i^-} M_i \exp\left(-\dfrac{\mu_i}{M_i}\dfrac{\delta x_i^-}{2}\right) \end{cases}
$$

These discretizations are slight modifications to the discretizations given in (2.3.7) and (2.3.10), respectively. For the planar SDE problems, we apply a 2D version of (3.1.2) given in (2.4.7). When the noise is additive, we use an evenly spaced grid. When the noise is multiplicative, we use adaptive mesh refinement as described in §3.4 in 1D and §3.8 in 2D. For the 39D SDE problem, we apply the generator on a gridless state space given in (2.5.2).

3.2. Cubic Oscillator in 1D with Additive Noise

This SDE problem is a concrete example of exploding numerical trajectories. Consider (2.1.2) with $\Omega = \mathbb{R}$, $\mu(x) = -x^3$, and $\sigma(x) = 1$ i.e.

$$dY = -Y^3 dt + \sqrt{2} dW \ , \quad Y(0) \in \mathbb{R}. \tag{3.2.1}$$

The solution to (3.2.1) is geometrically ergodic with respect to a stationary distribution with density

$$\nu(x) = Z^{-1} \exp(-x^4/4) \ , \quad \text{where } Z = \int_{\mathbb{R}} \exp(-x^4/4) dx. \tag{3.2.2}$$

However, explicit methods are transient in this example since the drift entering (3.2.1) is only locally Lipschitz continuous. More concretely, let $\{\tilde{X}_n\}$ denote a (discrete-time) Markov chain produced by forward Euler with time step size h_t. Then, for any $h_t > 0$ this Markov chain satisfies

$$\mathbb{E}_x \tilde{X}^2_{\lfloor t/h_t \rfloor} \to \infty \quad \text{as } t \to \infty \ ,$$

where $\mathbb{E}_x$ denotes expectation conditional on $\tilde{X}_0 = x$. (See Lemma 6.3 of [**64**] for a simple proof, and [**71**] for a generalization.) Metropolizing explicit integrators, like forward Euler, mitigates this divergence. Indeed, the (discrete-time) Markov chain produced by a Metropolis integrator $\{\bar{X}_n\}$ is ergodic with respect to the stationary distribution of the SDE, and hence,

$$\mathbb{E}_x \bar{X}^2_{\lfloor t/h_t \rfloor} \to \int_{\mathbb{R}} x^2 \nu(x) dx \quad \text{as } t \to \infty.$$

However, the rate of convergence of this chain to equilibrium is not geometric, even though the SDE solution is geometrically ergodic [**18, 136**]. The severity of this lack of geometric ergodicity was recently studied in [**15**], where it was shown that the deviation from geometric ergodicity is exponentially small in the time step size. This result, however, does not prevent the chain from "getting stuck" (rarely accepting proposal moves) in the tails of the stationary density in (3.2.2).

In this context we test the generators $\tilde{Q}_u$ in (3.1.1) and $\tilde{Q}_c$ in (3.1.2) on an infinite, evenly spaced grid on $\mathbb{R}$. Figure 3.1 plots sample paths produced by the SSA induced by these generators. The initial condition is large and positive. For comparison, we plot a sample path (in light grey) of a Metropolis integrator with the same initial condition. At this initial condition, the Metropolis integrator rarely accepts proposal moves and gets stuck for the duration of the simulation. In contrast, the figure shows that the proposed approximations do not get stuck, which is consistent with Theorem 4.2.8. The inset in the figure illustrates that the time lag between jumps in the SSA method is small (resp. moderate) at the tails (resp. middle) of the stationary density. This numerical result manifests that the mean holding time adapts to the size of the drift. Note that the time lag for $\tilde{Q}_c$ is smaller than for $\tilde{Q}_u$. In the next section, this difference in time lags is theoretically accounted for by an asymptotic analysis of the mean holding time. Figure 3.2 provides evidence that the approximations are accurate in representing the stationary distribution, mean first passage time, and committor function. To produce this figure, we use results described in Chapter 6. In particular, in the scalar case, the numerical stationary density, mean first passage time and committor satisfy a three-term recurrence relation, which can be exactly solved as detailed in Sections 6.1, 6.4 and 6.3, respectively. We also checked that these solutions agree – up to statistical error – with empirical estimates obtained by an SSA simulation.

REMARK 3.2.1. It is known that the rate of convergence (and hence accuracy) of the solution of standard time integrators to first exit problems drops because they fail to account for the probability that the system exits in between each discrete time step [**48**–**50**]. Building this capability into time integrators may require solving a boundary value problem per step, which is prohibitive to do in many dimensions [**113**]. By keeping time continuous, and provided that the grid is adjusted to fit the boundaries of the region of the first exit problem, the proposed approximations take these probabilities into account. As a consequence, in the middle panel of Figure 3.2 we observe that the order of accuracy of the approximations with respect to first exit times is identical to the accuracy of the generator, that is, there is no drop of accuracy as happens with time discretizations. To be precise, this figure illustrates the accuracy in these approximations with respect to the mean first passage time of the cubic oscillator to $(0,2)^c$ and shows that the generators $\tilde{Q}_u$ and $\tilde{Q}_c$ are $O(\delta x)$ and $O(\delta x^2)$, respectively.

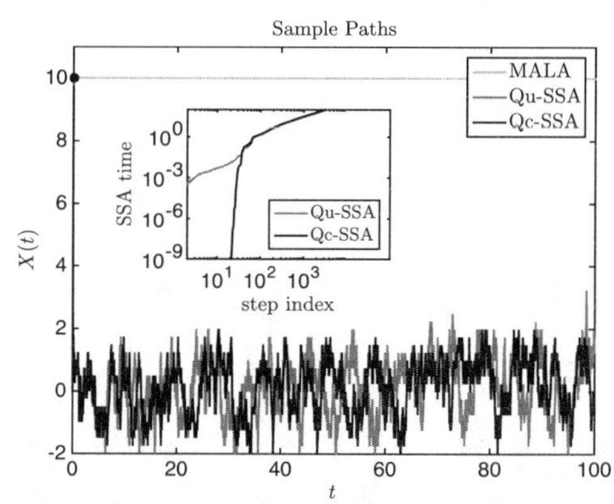

FIGURE 3.1. **Cubic Oscillator.** The main figure plots sample paths produced by a Metropolis integrator (MALA) and the SSA induced by $\tilde{Q}_u$ in (2.3.7) and $\tilde{Q}_c$ in (2.3.10) over the time interval $[0, 100]$. The temporal step size of MALA and the spatial step size of the SSA integrators are set at $\delta t = 0.015$ and $\delta x = 0.25$, respectively. The mean holding time for the SSA integrators operated at this spatial step size is approximately $\langle \delta t \rangle = 0.03$. At the initial condition marked by the black dot at $t = 0$, the true dynamics is dominated by the deterministic drift, which causes the true solution to drop toward the origin. Notice that MALA gets stuck at high energy, which illustrates that it is not geometrically ergodic. In contrast, SSA is geometrically ergodic (on an infinite grid), and hence, does not get stuck. The inset shows that the SSA time adapts to the size of the drift: when the drift is large (resp. small) the lag times are smaller (resp. larger).

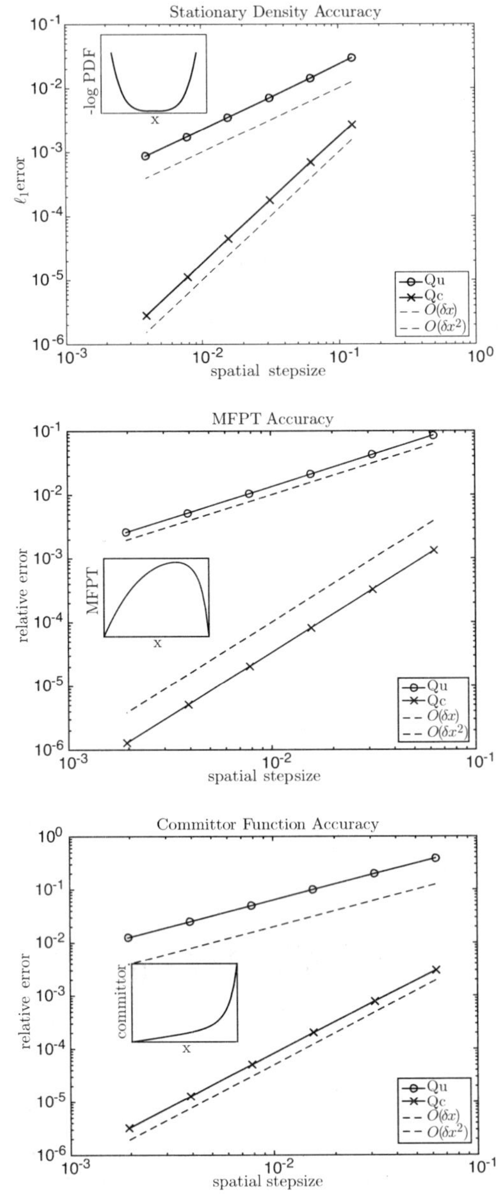

FIGURE 3.2. **Cubic Oscillator.** From top: accuracy of the generators $\tilde{Q}_u$ in (2.3.7) and $\tilde{Q}_c$ in (2.3.10) with respect to the stationary density, mean first passage time, and committor function of the SDE. A formula for the true solutions is explicitly available in this example and serves as a benchmark for comparison. For the numerical solution, we use the formulas given in Sections 6.1, 6.4 and 6.3. In all cases we see that $\tilde{Q}_u$ (resp. $\tilde{Q}_c$) is first (resp. second) order accurate in representing these quantities. The reference solutions are graphed in the insets.

3.3. Asymptotic Analysis of Mean Holding Time

Consider the following instance of (2.1.2)

$$dY = \mu(Y)dt + \sqrt{2}dW \ , \quad Y(0) \in \mathbb{R}. \tag{3.3.1}$$

Assume the drift entering (3.3.1) is differentiable and satisfies a dissipativity condition, like

$$\operatorname{sign}(x)\mu(x) \to -\infty \qquad \text{as } |x| \to \infty. \tag{3.3.2}$$

Hence, for sufficiently large $|x|$ the SDE dynamics is dominated by the drift, and asymptotically, the time it takes for the SDE solution $Y(t)$ to move a fixed distance in space can be derived from analyzing the ODE

$$\dot{Y} = \mu(Y) \ , \quad Y(0) = x. \tag{3.3.3}$$

Equation (3.3.3) implies that the time lapse between two grid points: $Y(0) = x_i$ and $Y(t^e) = x_{i-1}$ where $x_i > x_{i-1} \gg 0$ satisfies

$$t^e = \int_{x_{i-1}}^{x_i} \frac{dx}{|\mu(x)|}. \tag{3.3.4}$$

For simplicity, we assume that the spacing between grid points is uniform: $\delta x_i^+ = \delta x_i^- = \delta x$. By using integration by parts and the mean value theorem, observe that (3.3.4) can be written as:

$$\begin{aligned} t^e &= \frac{\delta x}{|\mu(x_i)|} - \int_{x_{i-1}}^{x_i} (x - x_{i-1})\mu(x)^{-2}\mu'(x)dx \\ &= t^* - \frac{1}{2}\frac{\mu'(\xi)}{\mu(\xi)^2}\delta x^2 \end{aligned} \tag{3.3.5}$$

for some $\xi \in (x_{i-1}, x_i)$, and where we have introduced: $t^* = \delta x / |\mu_i|$. Let us compare t^e to the mean holding times predicted by the generators: $\tilde{Q}_u$ in (3.1.1) and $\tilde{Q}_c$ in (3.1.2).

From (3.1.1) the mean holding time of $\tilde{Q}_u$ can be written as:

$$t^u = ((\tilde{Q}_u)_{i,i+1} + (\tilde{Q}_u)_{i,i-1})^{-1} = \frac{\delta x^2}{2 + |\mu_i|\delta x}$$

This expression can be rewritten as

$$t^u = t^* - t^* \frac{2}{2 + |\mu_i|\delta x}. \tag{3.3.6}$$

From (3.3.6), and (3.3.2), we see that t^u approaches t^* as $|x_i|$ becomes large, as the next Proposition states.

PROPOSITION 3.3.1. *For any $\delta x > 0$, the mean holding time of $\tilde{Q}_u$ satisfies:*

$$\frac{|t^u - t^*|}{t^*} \to 0 \quad \textit{as } |x_i| \to \infty.$$

Likewise, if the second term in (3.3.5) decays faster than the first term, then the relative error between t^e and t^* also tends to zero, and thus, the estimate predicted by $\tilde{Q}_u$ for the mean holding time asymptotically agrees with the exact mean holding time. The second term in (3.3.5) decays sufficiently fast if, e.g., the leading order term in $\mu(x)$ is of the form $-a\, x^{2p+1}$ for $p \geq 0$ and $a > 0$.

Repeating these steps for the mean holding time predicted by $\tilde{Q}_c$ yields:

$$t^c = ((\tilde{Q}_c)_{i,i+1} + (\tilde{Q}_c)_{i,i-1})^{-1} = \frac{\delta x^2}{2} \operatorname{sech}\left(\mu_i \frac{\delta x}{2}\right). \tag{3.3.7}$$

It follows from this expression that even though $\tilde{Q}_c$ is a second-order accurate approximation to L, it does not capture the right asymptotic mean holding time, as the next Proposition states.

PROPOSITION 3.3.2. *For any $\delta x > 0$, the mean holding time of $\tilde{Q}_c$ satisfies:*

$$\frac{|t^c - t^*|}{t^*} \to 1 \quad \text{as } |x_i| \to \infty.$$

Simply put, the mean holding time of $\tilde{Q}_c$ converges to zero too fast. This analysis is confirmed in the inset of Figure 3.3, which shows that the holding times for $\tilde{Q}_u$ asymptotically agree with the holding times of the generator Q_e that uses the exact mean holding time of Y. This generator was introduced in §2.3.3, see (2.3.16) for its definition. This numerical result agrees with the preceding asymptotic analysis. In contrast, note that the mean holding time of Q_c is asymptotically an underestimate of the exact mean holding time, as predicted by (3.3.7).

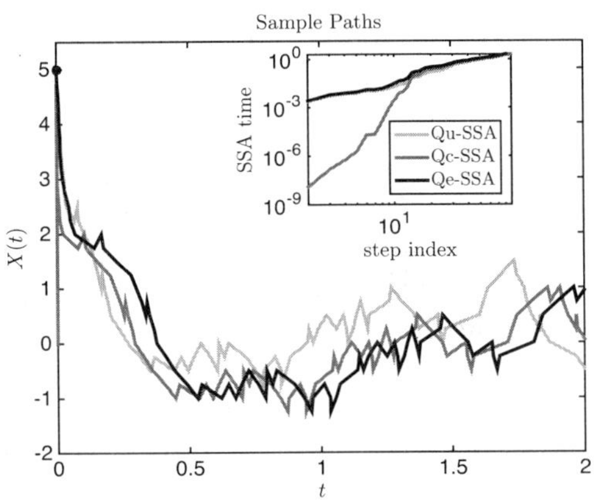

FIGURE 3.3. **Cubic Oscillator Simulation.** This figure plots paths produced by the SSA induced by $\tilde{Q}_u$ (light grey), $\tilde{Q}_c$ (grey) and Q_e (black). In all cases we assume a uniform spatial step size of $\delta x = 0.25$ and the initial condition is marked by the black dot at $t = 0$. The paths are driven by the same sequence of uniform random variables. We stress that the SSA produced by Q_e uses the exact mean holding time of the SDE solution, which far enough from the origin is well approximated by (3.3.5). The figure in the inset shows how the SSA time evolves with the SSA step index. Far away from the origin, the dynamics is dominated by the deterministic drift, which causes the walkers to move toward the origin and then become localized around the origin. The SSA associated to $\tilde{Q}_c$ moves to the origin too fast because its mean holding time tends to zero too fast, as (3.3.7) predicts. In contrast, the mean holding times for Q_u and Q_e are in close agreement far from the origin (where the asymptotic analysis in §3.3 is valid), but closer to the origin, this agreement is not apparent, as expected.

3.4. Adaptive Mesh Refinement in 1D

The subsequent scalar SDE problems have a domain $\Omega = \mathbb{R}^+$, and may have a singularity at the origin for certain parameter values. To efficiently resolve this singularity, we will use adaptive mesh refinement, and specifically, a variable step size infinite grid $S = \{x_i\} \in \mathbb{R}^+$ that contains more points near the singularity at $x = 0$. We will construct this grid by mapping the grid S to logspace to obtain a transformed grid $\hat{S}$ on $\mathbb{R}$ defined as

$$\hat{S} = \{\xi_i\} \, , \quad \xi_i = \log(x_i) \, , \quad \forall \, x_i \in S.$$

Note that as $\xi_i \to -\infty$ (resp. $\xi_i \to \infty$) we have that $x_i \to 0$ (resp. $x_i \to \infty$). For the transformed grid in logspace, we assume that the distance between neighboring grid points is fixed and given by $\delta\xi$, i.e.,

$$\delta\xi = \xi_{i+1} - \xi_i \quad \text{for all } i \in \mathbb{Z}.$$

Since

$$x_{i+1} = \exp(\xi_{i+1}) = \exp(\delta\xi)\exp(\xi_i) = \exp(\delta\xi)x_i$$

it follows from the definitions introduced in (2.3.5) that:

$$\delta x_i^+ = (\exp(\delta\xi) - 1)x_i \, , \quad \delta x_i^- = (1 - \exp(-\delta\xi))x_i \, , \quad \delta x_i = \sinh(\delta\xi)x_i. \tag{3.4.1}$$

3.5. Log-normal Process in 1D with Multiplicative Noise

Consider (2.1.2) with

$$\Omega = \mathbb{R}^+ \, , \quad \mu(x) = -x\log x + x \, , \quad M(x) = (\sigma(x))^2 = x^2 \, ,$$

and initial condition $Y(0) \in \Omega$. This process has a lognormal stationary distribution with probability density:

$$\nu(x) = \frac{1}{\sqrt{2\pi}} \exp\left(-\frac{1}{2}(\log(x))^2 - \log(x)\right). \tag{3.5.1}$$

In fact, $\log(Y(t))$ satisfies an Ornstein-Uhlenbeck equation with initial condition $\log(Y(0))$. Exponentiating the solution to this Ornstein-Uhlenbeck equation yields

$$Y(t) = \exp\left(e^{-t}\log Y(0) + \sqrt{2}\int_0^t e^{-(t-s)}dW(s)\right).$$

It follows that:

$$\mathbb{E}_x(Y(t)^2) = \exp\left(2e^{-t}\log(x) + 2(1 - e^{-2t})\right). \tag{3.5.2}$$

The formulas (3.5.1) and (3.5.2) are useful to numerically validate the approximations.

Figure 3.4 show numerical results of applying the generators in (3.1.1) and (3.1.2) to solve the boundary value problems associated to the mean first passage time and exit probability (or committor function) on the interval $[1/2, 5]$ and using an evenly spaced mesh. Figure 3.5 illustrates that these generators are accurate with respect to the stationary density (3.5.1) using the adaptive mesh refinement described in §3.4. Figure 3.6 plots sample paths produced by SSA for these generators and the SSA time as a function of the step index. Figure 3.7 illustrates the accuracy of the proposed approximations with respect to the mean-squared displacement $\mathbb{E}_x(Y(1)^2)$ with an initial condition of $Y(0) = x = 2$. The relative error is plotted as both a function of the spatial step size (top panel) and the mean lag time (bottom panel). Finally, Figure 3.8 plots the average number of computational steps (top panel) and the mean lag time (bottom panel) as a function of the spatial step size. The statistics are obtained by averaging this data over 100 SSA realizations with initial condition $X(0) = 1$ and time interval of simulation of $[0, 10]$.

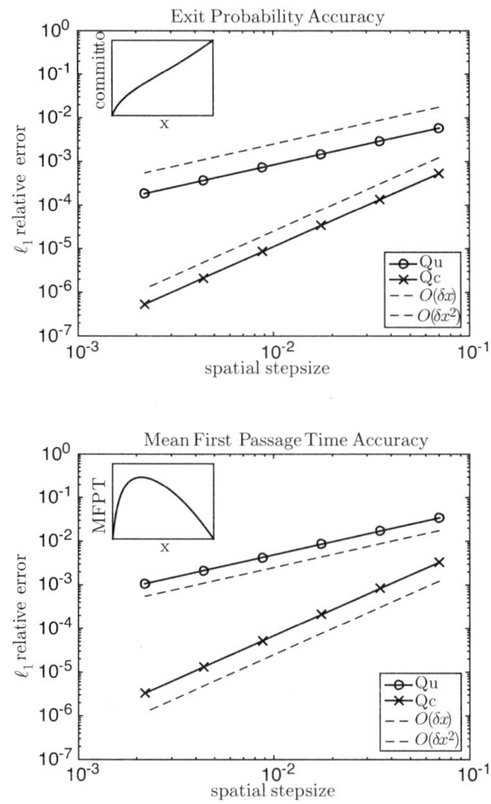

FIGURE 3.4. **Log-normal Process in 1D.** This figure confirms that the realizable discretizations given in (3.1.1) and (3.1.2) are accurate with respect to the mean first passage time (bottom panel) and exit probability or committor function (top panel) computed on the interval $[1/2, 5]$. The reference solution is shown in the inset. An evenly spaced mesh is used to produce this numerical result, with spatial step sizes as indicated in the figures. The figure confirms that $\tilde{Q}_u$ (resp. $\tilde{Q}_c$) is first (resp. second) order accurate.

3.6. Cox-Ingersoll-Ross Process in 1D with Multiplicative Noise

Consider a scalar SDE problem with boundary

$$dY = \beta(\alpha - Y)dt + \sigma\sqrt{Y}dW \,, \quad Y(0) \in \mathbb{R}^+ . \tag{3.6.1}$$

In quantitative finance, this example is a simple model for short-term interest rates [**30**]. It is known that discrete-time integrators may not satisfy the non-negativity condition: $Y(t) \geq 0$ for all $t > 0$ [**5**, **106**]. Corrections to these integrators that

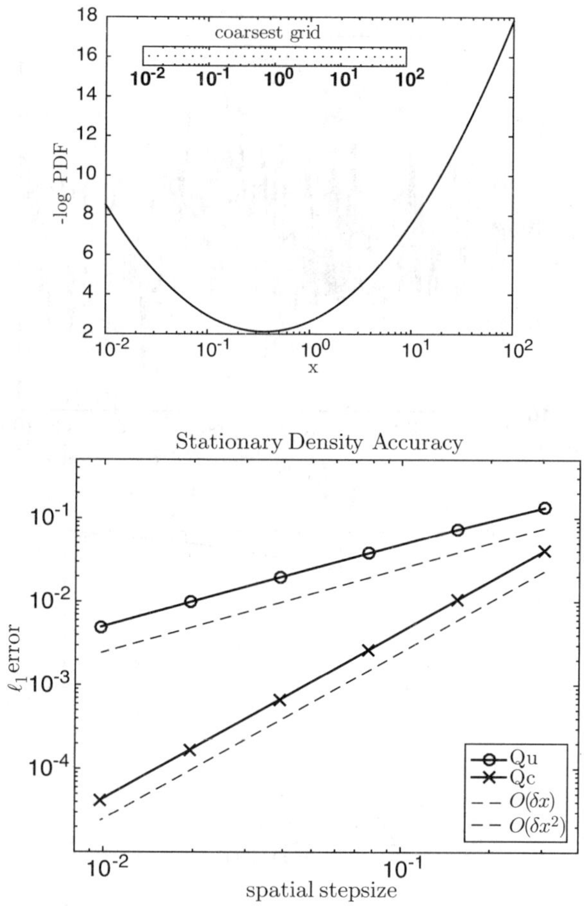

FIGURE 3.5. **Log-normal Process in 1D.** The top panel plots the free energy of the stationary density of the SDE. The bottom panel plots the accuracy of $\tilde{Q}_u$ and $\tilde{Q}_c$ in representing this stationary density as a function of the spatial step size $\delta\xi$ in log-space. The benchmark solution is the stationary density (3.5.1) evaluated on the variable step size grid described in §3.4. Note that the central scheme $\tilde{Q}_c$ appears to be second-order accurate in representing the stationary probability distribution of the SDE, which numerically supports Theorem 4.5.2.

enforce this condition tend to be involved, specific to this example, or lower the accuracy of the approximation. This issue motivates applying realizable discretizations to this problem, which can preserve non-negativity by construction.

According to the Feller criteria, the boundary at zero is classified as:

- natural if $2\beta\alpha/\sigma^2 \geq 1$;
- regular if $1 > 2\beta\alpha/\sigma^2 > 0$; and,
- absorbing if $\beta\alpha < 0$.

When the boundary at zero is natural (resp. regular), the process is positive (resp. non-negative). When the origin is attainable, the boundary condition at the origin is treated as reflective. If $\beta > 0$ and $\alpha > 0$, the SDE solution is ergodic with respect

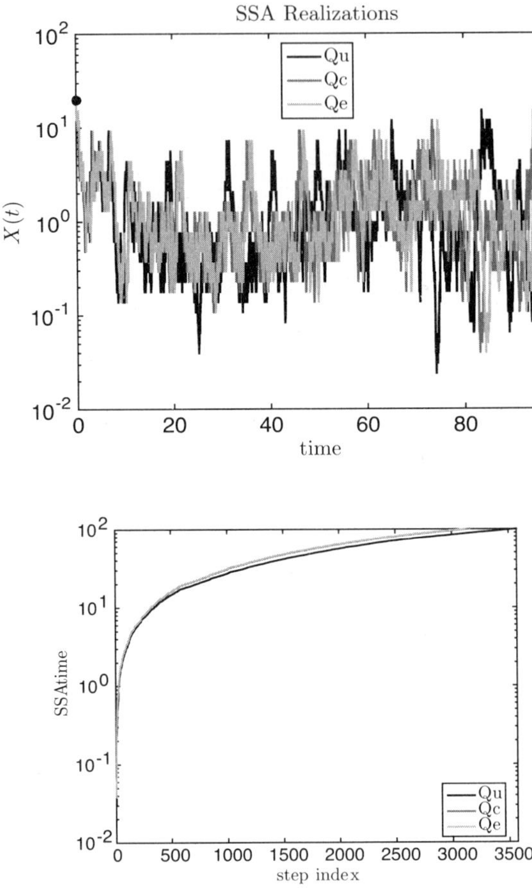

FIGURE 3.6. **Log-Normal Process in 1D.** The top panel plots realizations of the generators $\tilde{Q}_u$, $\tilde{Q}_c$, and Q_e produced using the SSA integrator with the initial condition at $X(0) = 20$ marked by the black dot at $t = 0$. Recall that Q_e reproduces the true mean holding time of the process, see (2.3.16) for its definition. The state space of the processes is the variable step size grid described in §3.4 with a grid size in log-space of $\delta\xi = 0.25$. The time interval of simulation is $[0, 100]$. The mean SSA time for all three methods is approximately 0.03. The bottom panel plots the SSA time as a function of the computational step index. This figure illustrates that the SSA integrators never exit the domain of definition of the coefficients of the SDE.

to a stationary distribution with density proportional to:

$$\nu(x) = x^{\frac{2\beta\alpha}{\sigma^2}-1} \exp\left(-\frac{2\beta}{\sigma^2}x\right). \tag{3.6.2}$$

We emphasize that $\nu(x)$ is integrable over $\mathbb{R}^+$ if and only if $\beta > 0$ and $\alpha > 0$. Note that when the boundary at zero is regular, the stationary density in (3.6.2) is unbounded at the origin.

The numerical tests assess how well the approximation captures the true long-time behavior of the SDE. To carry out this assessment, we estimate the difference

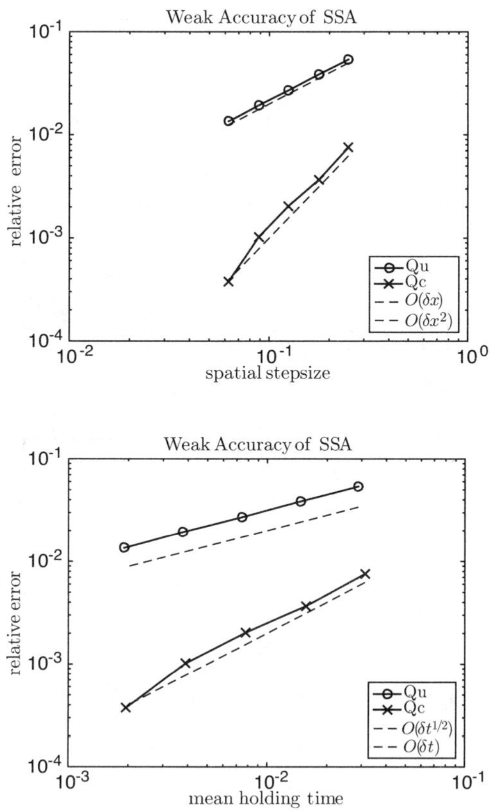

FIGURE 3.7. **Log-Normal Process in 1D.** The top (resp. bottom) panel plots the accuracy of the generators $\tilde{Q}_u$ and $\tilde{Q}_c$ with respect to $\mathbb{E}_x(Y(t)^2)$ as a function of spatial step size (resp. mean holding time) with initial condition $Y(0) = 2$ and terminal time $t = 1$. Note that the central scheme $\tilde{Q}_c$ appears to be second-order accurate, which numerically supports Theorem 4.5.1. This figure was produced using 10^8 SSA realizations.

between the numerical stationary densities and

$$\bar{\nu}_i = Z^{-1} \int_{\frac{x_i + x_{i-1}}{2}}^{\frac{x_i + x_{i+1}}{2}} \nu(x) dx \quad \text{for all } x_i \in S \tag{3.6.3}$$

where the constant Z is chosen so that $\sum_i \bar{\nu}_i = 1$. Figures 3.9 and 3.10 illustrate the results of using adaptive mesh refinement. The top panels of Figures 3.9 and 3.10 plot the free energy of the stationary density when the boundary at zero is natural and regular, respectively. The benchmark solution is the cell-averaged stationary density (3.6.2) on the adaptive mesh shown in the inset. Note that when the boundary at zero is natural, the minimum of the free energy of the stationary density is centered away from zero. In contrast, when the boundary is regular, the minimum of the free energy is attained at the grid point closest to zero. The bottom panels of Figures 3.9 and 3.10 plot the ℓ_1 error in the numerical stationary densities

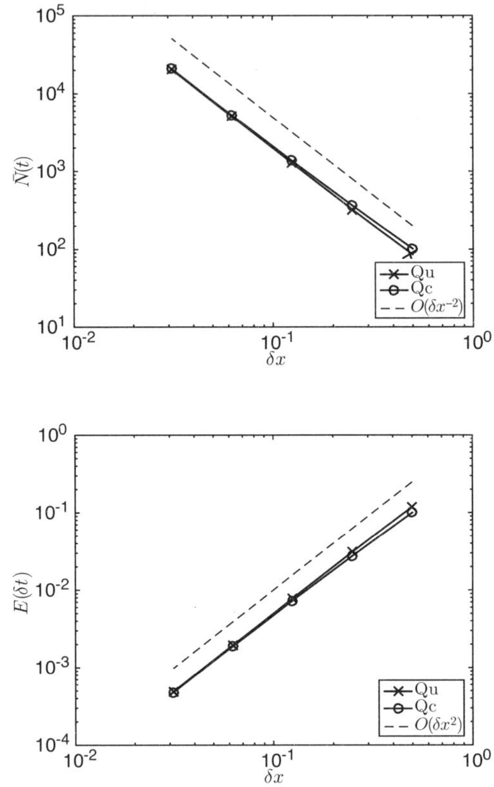

FIGURE 3.8. **Log-Normal Process in 1D.** These statistics are produced using the SSA induced by the generators $\tilde{Q}_u$ in (3.1.1) and $\tilde{Q}_c$ in (3.1.2) as indicated in the figure Legend. The top panel of this figure plots the mean number of computational steps of the SSA integrator as a function of the spatial step size. The bottom panel of this figure plots the mean holding time of the SSA integrator as a function of the spatial step size. The statistics are obtained by taking a sample average over 100 realizations with initial condition at $X(0) = 1$ and a time interval of simulation of $[0, 10]$. This figure validates Proposition 4.3.11.

as a function of the mesh size in logspace $\delta\xi$. These numerical results confirm the theoretically expected rates of convergence when the mesh size is variable.

3.7. SDEs in 2D with Additive Noise

The examples in this part are two-dimensional SDEs of the form

$$dY = b(Y)dt - DU(Y)dt + \sqrt{2\beta^{-1}}dW \ , \quad Y(0) \in \Omega \subset \mathbb{R}^2 \ , \tag{3.7.1}$$

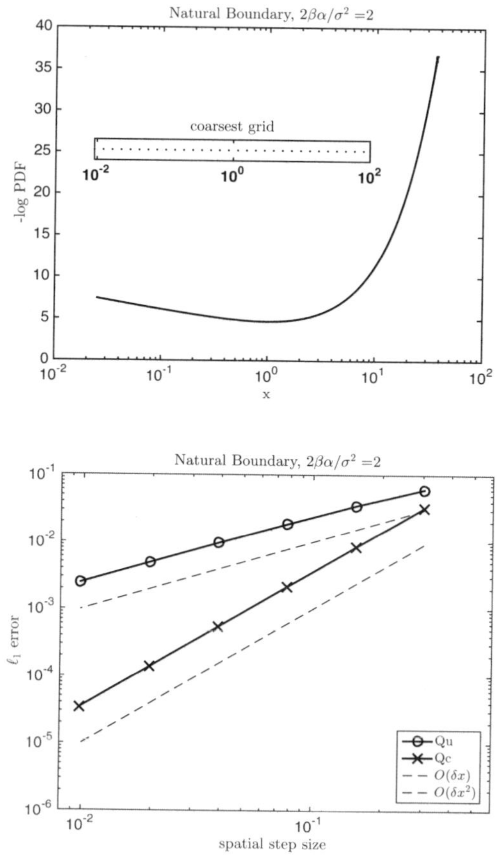

FIGURE 3.9. **Cox-Ingersoll-Ross Simulation.** The top panel plots the free energy of the stationary density of the SDE, when the boundary is natural. The bottom panel plots the accuracy of $\tilde{Q}_u$ and $\tilde{Q}_c$ in representing this stationary density as a function of the spatial step size $\delta\xi$ in log-space. The benchmark solution is the cell averaged stationary density (3.6.3) using the variable step size grid described in §3.4. Note that the central scheme $\tilde{Q}_c$ is second-order accurate in representing the stationary probability distribution of the SDE, which numerically supports Theorem 4.5.2.

where $U : \mathbb{R}^2 \to \mathbb{R}$ is a potential energy function (which will vary for each example), β is the inverse 'temperature', and $b(x)$ is one of the following flow fields:

$$(3.7.2) \qquad b(x) \in \left\{ \underbrace{\begin{pmatrix} \gamma x_2 \\ -\gamma x_1 \end{pmatrix}}_{\text{rotational}}, \underbrace{\begin{pmatrix} \gamma x_2 \\ \gamma x_1 \end{pmatrix}}_{\text{extensional}}, \underbrace{\begin{pmatrix} \gamma x_2 \\ 0 \end{pmatrix}}_{\text{shear}}, \underbrace{\begin{pmatrix} -\gamma x_1 x_2^2 \\ 0 \end{pmatrix}}_{\text{nonlinear}} \right\}, \quad x = (x_1, x_2)^T$$

where γ is a flow strength parameter. In the flow-free case ($\gamma = 0$) and in the examples we consider, the generator of the SDE in (3.7.1) is self-adjoint with respect

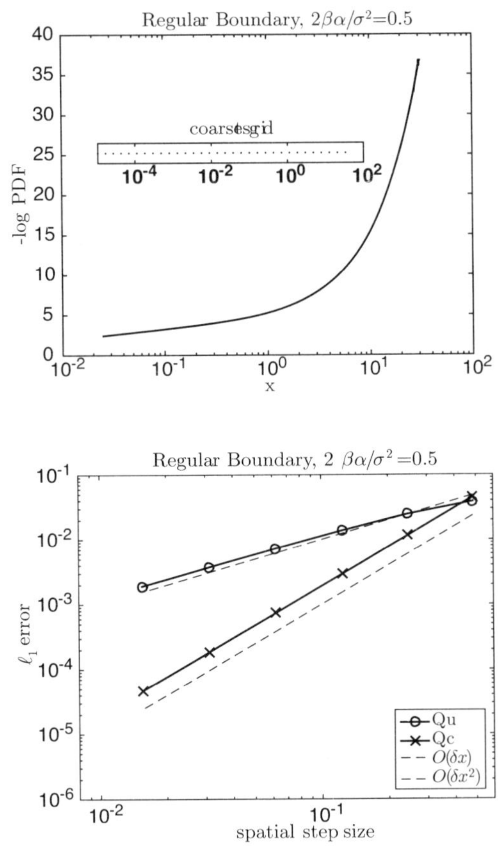

FIGURE 3.10. **Cox-Ingersoll-Ross Simulation.** The top panel plots the free energy of the stationary density of the SDE, when the boundary is regular. The bottom panel plots the accuracy of $\tilde{Q}_u$ and $\tilde{Q}_c$ in representing this stationary density as a function of the spatial step size parameter $\delta\xi$ in log-space. The benchmark solution is the cell averaged stationary density (3.6.3) using the variable step size grid described in §3.4. Note that the central scheme $\tilde{Q}_c$ is second-order accurate in representing the stationary probability distribution of the SDE, which numerically supports Theorem 4.5.2.

to the probability density:

$$\nu(x) = Z^{-1} \exp(-\beta U(x)) \ , \quad Z = \int_\Omega \exp(-\beta U(x)) dx.$$

Moreover, the process Y is reversible [**58**]. Realizable discretizations have been developed for such diffusions with additive noise in arbitrary dimensions, see [**102**, **121**]. With flow ($\gamma \neq 0$), however, the generator is no longer self-adjoint, the equilibrium probability current is nonzero, and the stationary density of this SDE is no longer $\nu(x)$, and in fact, is not explicitly known. In this context we will assess how well the second-order accurate generator $\tilde{Q}_c$ in (2.4.7) reproduces the spectrum of the infinitesimal generator of the SDE L. Since the noise is additive, the state space of the approximation given by $\tilde{Q}_c$ is a grid over $\Omega \subset \mathbb{R}^2$. Moreover, this generator can be represented as an infinite-dimensional matrix. In order to use sparse matrix eigensolvers to compute eigenvalues of largest real part and the leading left eigenfunction of $\tilde{Q}_c$, we will approximate this infinite-dimensional matrix using a finite-dimensional matrix. For this purpose, we truncate the grid as discussed next.

REMARK 3.7.1 (Pruned Grid). Let $\{x_{ij} \in \mathbb{R}^2 \mid (i,j) \in \mathbb{Z}^2\}$ be the set of all grid points and consider the magnitude of the drift field evaluated at these grid points: $\{|\mu(x_{ij})|\}$. In order to truncate the infinite-dimensional matrix associated to $\tilde{Q}_c$, grid points at which the magnitude of the drift field is above a critical value $E^\star$ are pruned from the grid, and the jump rates to these grid points are set equal to zero. The price of this type of truncation is an unstructured grid that requires using an additional data structure to store a list of neighbors to each unpruned grid point. To control errors introduced in this pruning, we verify that computed quantities do not depend on the parameter $E^\star$.

REMARK 3.7.2 (Error Estimates). In the linear context, we compare to the exact analytical solution. In the nonlinear context, and in the absence of an analytical solution, we measure error relative to a converged numerical solution.

3.7.1. Planar, Asymmetric Ornstein-Uhlenbeck Process. Consider (3.7.1) with the following two dimensional potential energy:

$$U(x) = \frac{1}{2}(x_1^2 + x_2^2)\ , \qquad x = (x_1, x_2)^T \in \mathbb{R}^2\ , \tag{3.7.3}$$

and the linear flows defined in (3.7.2). We will use this test problem to assess the accuracy of the generator $\tilde{Q}_c$ in representing the spectrum of L.

In the linear context (3.7.1) with $\beta = 2$, the SDE (3.7.1) can be written as:

$$dY = CY dt + dW \tag{3.7.4}$$

where C is a 2×2 matrix. The associated generator L is a two-dimensional, non-symmetric Ornstein-Uhlenbeck operator. Let λ_1 and λ_2 denote the eigenvalues of the matrix C. If $\Re(\lambda_1) < 0$ and $\Re(\lambda_2) < 0$, then L admits a unique Gaussian stationary density $\nu(x)$ [**72**, See Section 4.8], and in addition, the spectrum of L in L^p_ν (for $1 < p < \infty$) is given by:

$$\sigma(L) = \{n_1\lambda_1 + n_2\lambda_2 \ : \ n_1, n_2 \in \mathbb{N}_0\}. \tag{3.7.5}$$

The original statement and proof of (3.7.5) in arbitrary, but finite, dimension can be found in [**119**]. This result serves as a benchmark for the approximation. In the numerical tests, we select the flow rate to be $\gamma = 1/2$, which ensures that the eigenvalues of the matrix C have strictly negative real parts for each of the linear flow fields considered.

Figure 3.11 plots contours of the free energy of the numerical stationary density at $\delta x = 0.3$ for the linear flow fields considered. At this spatial step, the ℓ_1 error in the stationary densities shown is about 1%. Figure 3.12 plots the ℓ_1 difference between the numerical and exact stationary densities. Figure 3.13 plots the first twenty eigenvalues of L as predicted by (3.7.5). Figure 3.14 plots the relative error between the twenty eigenvalues of largest real part of L and Q_c. The eigenvalues of L are obtained from (3.7.5), and the eigenvalues of $\tilde{Q}_c$ are obtained from using a sparse matrix eigensolver and the truncation of $\tilde{Q}_c$ described in Remark 3.7.1.

This last figure suggests that the accuracy of $\tilde{Q}_c$ in representing the spectrum of L is, in general, first-order.

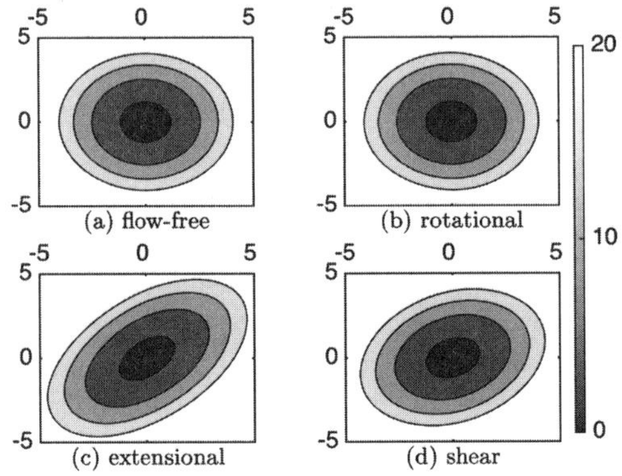

FIGURE 3.11. **Planar, Asymmetric Ornstein-Uhlenbeck Process.** These figures plot contours of the free energy of the numerical stationary density associated to $\tilde{Q}_c$ in the flow-free (a), rotational (b), extensional (c), and shear flow (d) cases. The grid size is selected to be $h = 0.3$, which yields an approximation to the true stationary density that is within 1% error in the ℓ_1 norm. The circular contour lines of the stationary densities in the (a) flow-free and (b) rotational cases are identical. In the (c) extensional and (d) shear flow cases, the contours are elliptical. The flow strength parameter is set at: $\gamma = 1/2$.

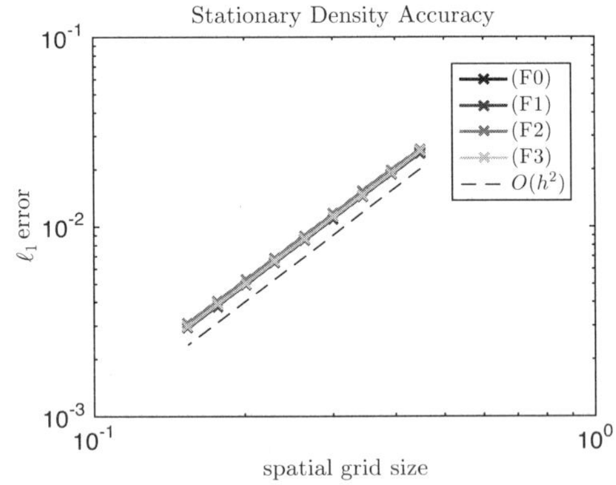

FIGURE 3.12. **Planar, Asymmetric Ornstein-Uhlenbeck Process.** These figures plot the ℓ_1 error in the estimate of the true stationary density as a function of grid size for $\tilde{Q}_c$ in the flow-free (F0), rotational (F1), extensional (F2), and shear (F3) flow cases, as indicated in the figure legend. The figure shows that this approximation is second-order accurate in estimating the stationary density. The flow strength parameter is set at: $\gamma = 1/2$. This numerical result is consistent with Theorem 4.5.2.

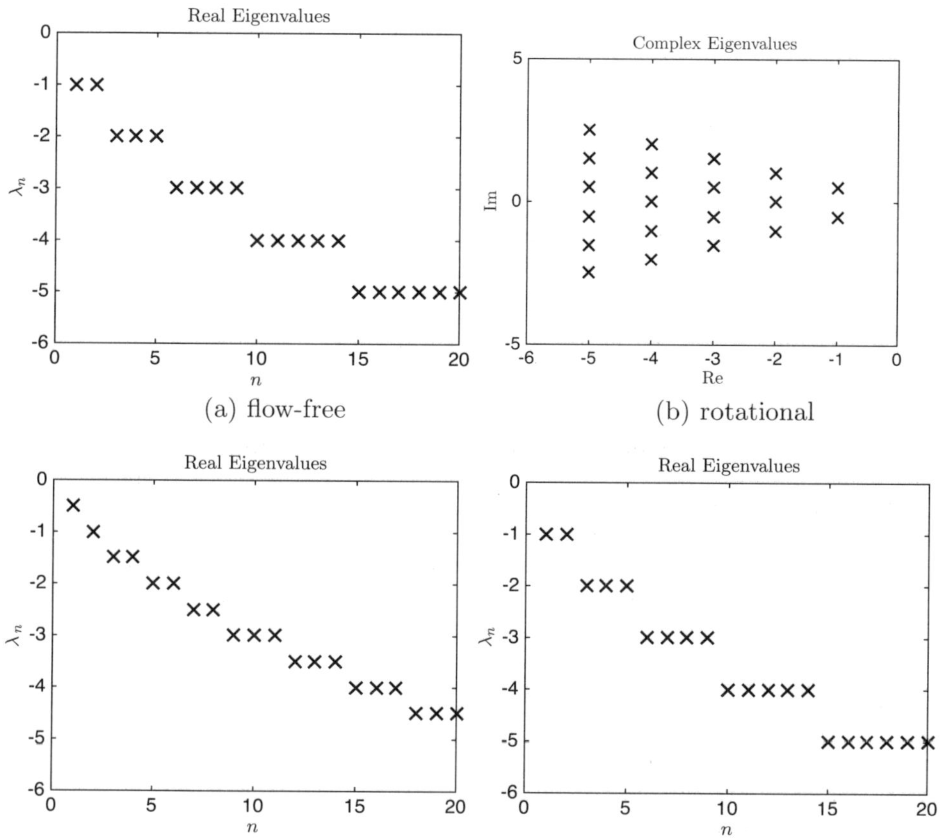

(a) flow-free (b) rotational

(c) extensional (d) shear

FIGURE 3.13. **Planar, Asymmetric Ornstein-Uhlenbeck Process.** These figures plot the twenty eigenvalues of largest nontrivial real part of the operator L as predicted by the formula given in (3.7.5). In (a), (c) and (d) the formula implies that the eigenvalues are strictly real, while in the rotational flow case shown in (b) the eigenvalues are complex. The flow strength parameter is set at: $\gamma = 1/2$. These eigenvalues serve as a benchmark solution to validate the accuracy of $\tilde{Q}_c$ in representing the spectrum of L.

3.7.2. Maier-Stein SDE: Locally Lipschitz Drift & Nonlinear Flow. The Maier-Stein system is governed by an SDE of the form (3.7.1) with $\Omega = \mathbb{R}^2$, the nonlinear flow field in (3.7.2), $\beta = 1$, and

$$U(x) = \frac{x_1^4}{4} - \frac{x_1^2}{2} + \mu\frac{(1+x_1^2)x_2^2}{2}, \qquad x = (x_1, x_2)^T \in \mathbb{R}^2. \tag{3.7.6}$$

In the flow-free case ($\gamma = 0$), the Maier-Stein SDE describes the evolution of a self-adjoint diffusion. In this case, the governing SDE is an overdamped Langevin equation with a drift that is the gradient of a double well potential with two minima located at the points $(\pm 1, 0)^T$. When $\gamma > 0$, this process is no longer self-adjoint and the drift cannot be written as the gradient of a potential [**109–111**]. Our interest in this system is to verify that (2.4.7) is convergent for SDEs of the type

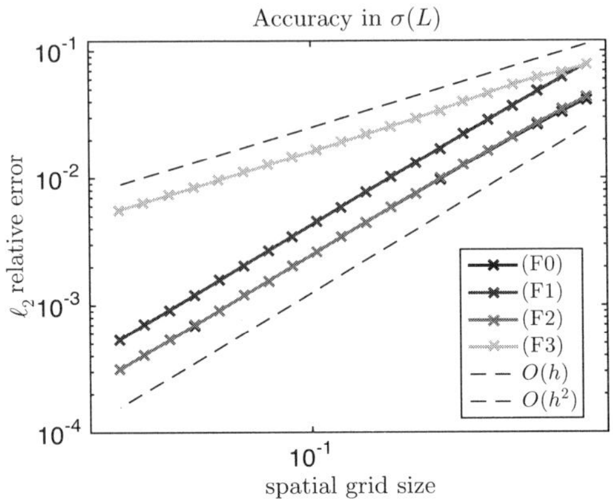

FIGURE 3.14. **Planar, Asymmetric Ornstein-Uhlenbeck Process.** These figures plot the relative error of the numerical estimate produced by $\tilde{Q}_c$ of the twenty eigenvalues of largest nontrivial real part of the operator L in the flow-free (F0), rotational (F1), extensional (F2), and shear (F3) flow cases, as indicated in the figure legend. The figures suggest that the accuracy of $\tilde{Q}_c$ in representing these eigenvalues is $\mathcal{O}(h)$ in general, even though it is $\mathcal{O}(h^2)$ in the flow-free (F0), rotational (F1), and extensional (F2) flow cases. The flow strength parameter is set at: $\gamma = 1/2$.

(3.7.1) with a nonlinear flow field. Figure 3.15 reproduces slices of the stationary density, which is consistent with Figure 5 of [**155**]. Contours of the free energy of the stationary density at $\gamma = 0$ and $\gamma = 4$ are shown in Figure 3.16. Finally, Figure 3.17 shows that the spectrum of L is accurately represented by $\tilde{Q}_c$.

3.7.3. Square Well SDE: Discontinuous Potential. Here, we test $\tilde{Q}_c$ in (2.4.7) on a non-self-adjoint diffusion whose coefficients have internal discontinuities. Specifically, consider (3.7.1) with the linear flow fields in (3.7.2), $\beta = 1$, and a regularized square well potential:

$$
\begin{aligned}
U(x) &= \tanh\left(\frac{\max(x_1, x_2) - d_1}{\epsilon}\right) - \tanh\left(\frac{\max(x_1, x_2) - d_2}{\epsilon}\right) \\
x &= (x_1, x_2)^T \in \Omega = [-1, 1] \times [-1, 1]
\end{aligned}
\tag{3.7.7}
$$

where d_1 and d_2 specify the location of the discontinuity and ϵ is a smoothness parameter. We assume that the solution is periodic at the boundaries of Ω and we discretize the domain so that all of the grid points are in the interior of Ω. Since the potential energy is nearly discontinuous, we replace the jump rates in (2.5.2) with the following rates which avoid differentiating the potential energy

$$
q(x, y) = \frac{\beta^{-1}}{h^2} \exp\left(-\frac{\beta}{2}(U(y) - U(x) - b(x)^T (y - x))\right). \tag{3.7.8}
$$

To be clear, x, y in (3.7.8) are 2-vectors. Contours of the numerical stationary density are plotted in Figure 3.18 when the stationary density is known. The spatial grid size used is $h = 0.1$, which is selected so that the numerical densities represent the true stationary densities to within 1% accuracy in the ℓ_1 norm. Figure 3.19 shows a similar plot in the flow cases for which the stationary densities are unknown.

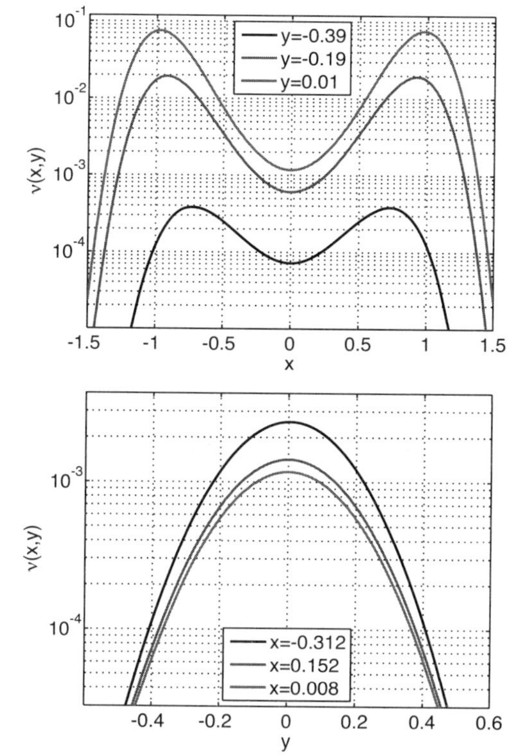

FIGURE 3.15. **Maier-Stein: Slices of Stationary Density.** These figures plot the intersection of the stationary density of the Maier-Stein system with the planes indicated in the legends with parameters: $\mu = 2$ and $\gamma = 4.67$. These graphs agree with those given in Figure 5 of [**155**].

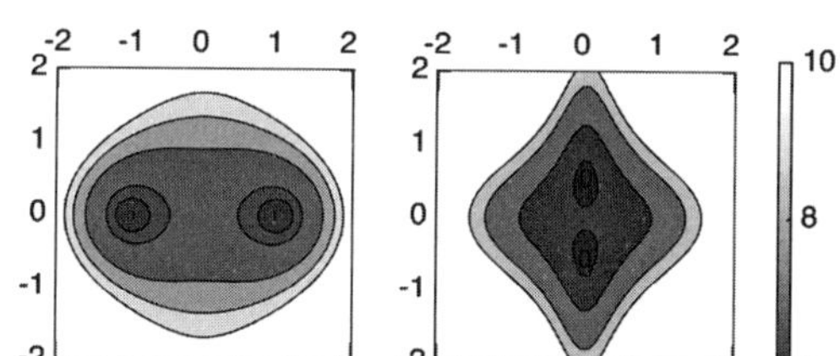

(a) flow-free (b) nonlinear flow

FIGURE 3.16. **Maier-Stein: Stationary Densities.** These figures plot contours of the free energy of the stationary density for the generator $\tilde{Q}_c$ with grid size $h = 0.05$. The test problem is a particle in the double-well potential given in (3.7.6) and the nonlinear flow field given in (3.7.2). The flow strength parameters are set at: $\gamma = 0$ (left panel) and $\gamma = 4$ (right panel).

In this case we checked that at the spatial grid size of $h = 0.1$ the stationary density is well-resolved. To accurately represent the probability flux through the square well, we construct the coarsest cell-centered mesh to have borderlines that coincide with the discontinuities of the square well. Subsequent refinements involve

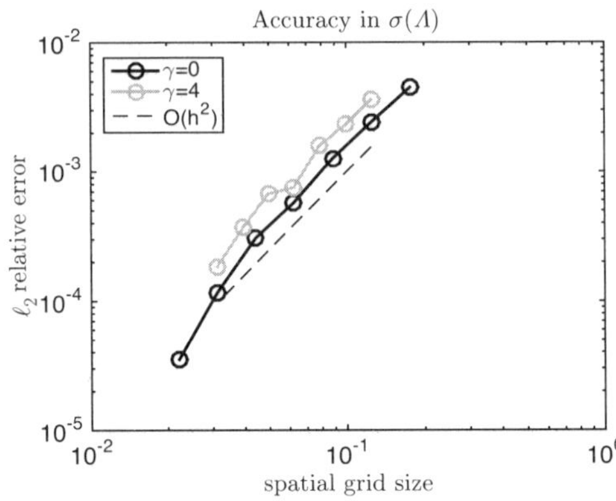

FIGURE 3.17. **Maier-Stein: Accuracy in $\sigma(L)$.** This figure illustrates the convergence in the ℓ_2 norm of the numerical estimate produced by $\tilde{Q}_c$ of the twenty eigenvalues of largest nontrivial real part of the operator L for the values of the flow strength parameter γ indicated in the legend.

halving the size of each cell, so that the discontinuities always lie on a cell edge. Figure 3.20 illustrates the ℓ_2 convergence of the numerical estimate to the first twenty eigenvalues of the continuous operator L. As a side note, one may be able to construct a more accurate generator in this context by extending the spatial discretization introduced in [**159**] to non-self-adjoint diffusions.

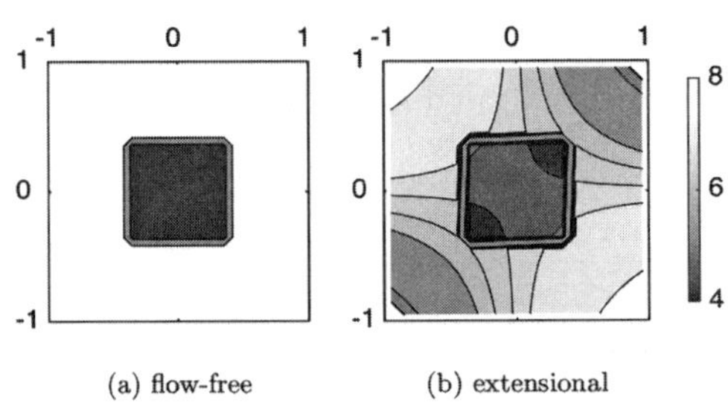

FIGURE 3.18. **Square Well: Stationary Densities (known).** These figures plot contours of the free energy of the numerical stationary density for a particle in the potential given by (3.7.7) with $\beta = 1$, $d_1 = 0.4$, $d_2 = -0.4$, $\epsilon = 0.001$, and the flow-free (left panel) and extensional flow (right panel) cases with $\gamma = 2$. For these flows a formula for the stationary density $\nu(x)$ is available, and the numerical densities shown are within 1% accurate in the $\ell_1$norm.

3.7.4. Worm-Like Chain SDE: Singular Drift. Consider (3.7.1) with the linear flow fields in (3.7.2), $\beta = 1$, and the potential energy given by:

$$(3.7.9)\quad U(x) = \frac{x_2^2}{2} + \frac{1}{1-|x_1|} - |x_1| + 2x_1^2\,, \qquad x = (x_1, x_2)^T \in \Omega = (-1,1) \times \mathbb{R}\,,$$

This system is an example of an SDE with a boundary and with a drift vector field that has singularities at the boundaries of Ω: $(\pm 1, x_2)^T$ for all $x_2 \in \mathbb{R}$. Because

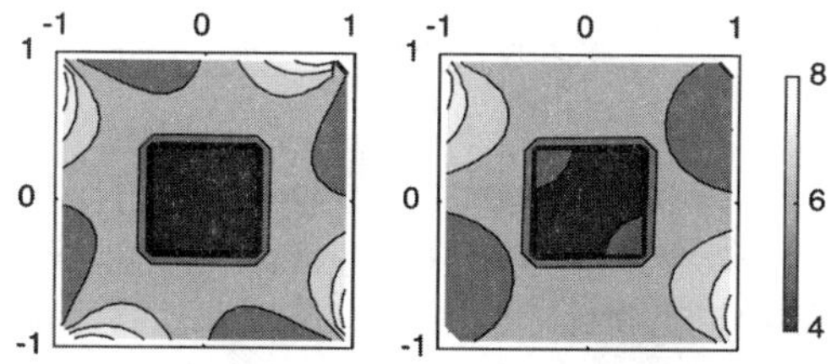

(a) rotational (b) shear

FIGURE 3.19. **Square Well: Stationary Densities (unknown).** These figures plot contours of the free energy of the numerical stationary density for a particle in the potential given by (3.7.7) with $\beta = 1$, $d_1 = 0.4$, $d_2 = -0.4$, $\epsilon = 0.001$, and the shear (left panel) and rotational flow (right panel) cases with $\gamma = 2$. For these flows a formula for the stationary density $\nu(x)$ is not available, but we verified that these numerical solutions are well-resolved

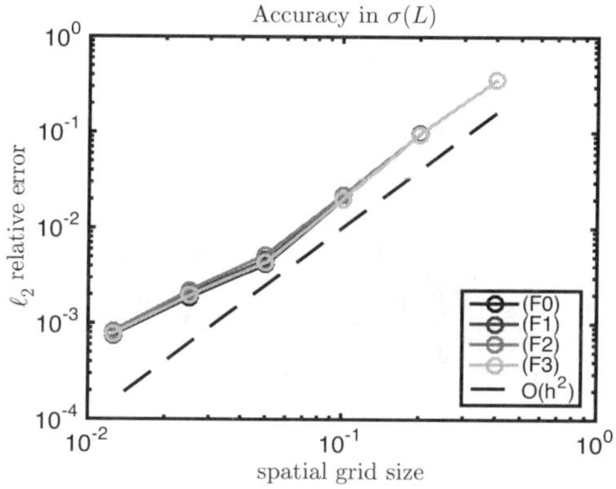

FIGURE 3.20. **Square Well: Accuracy in $\sigma(L)$.** This figure plots the relative error in the ℓ_2 norm of the numerical estimate produced by $\tilde{Q}_c$ of the twenty eigenvalues of largest nontrivial real part of the operator L. The parameters here are set at: $\gamma = 1$ and $\beta = 1$. The figure suggests that the relative error of $\tilde{Q}_c$ in representing the spectrum of L is about second-order in the spatial step size. The slight divergence at small spatial step size is attributed to errors in the sparse matrix eigensolver used.

of this singularity, standard integrators for (3.7.1) may diverge in finite time even in the flow-free case. Contour lines of the numerical stationary density $\nu(x)$ are plotted in Figure 3.21 and 3.22 when the stationary density is known and unknown, respectively, and for a flow rate of $\gamma = 5$. We choose a high flow rate to make the numerical test more rigorous. The accuracy of the spectrum of the generator $\tilde{Q}_c$ is assessed in Figure 3.23. This figure suggests that $\tilde{Q}_c$ is second-order accurate in representing the spectrum of the generator L.

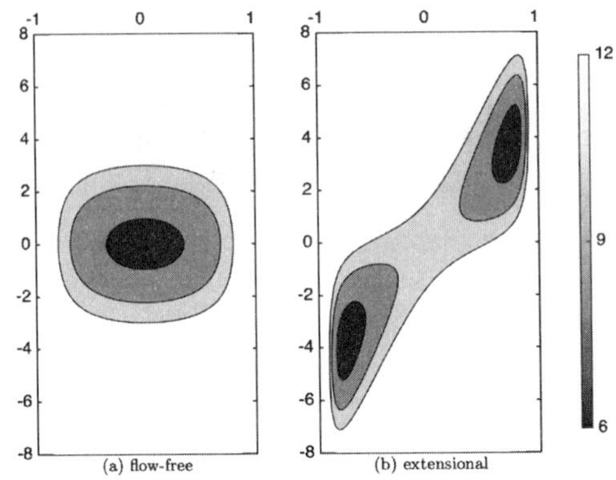

FIGURE 3.21. **Worm-Like Chain: Stationary Densities (Known).** These figures plot contours of the stationary density for a particle in the potential given by (3.7.9), and from left, flow-free and the extensional flow given in (3.7.2) with $\gamma = 5$ and $\beta = 1$. For these flows a formula for the stationary density $\nu(x)$ is available, and the numerical densities shown are within 1% accurate in the $\ell_1$norm.

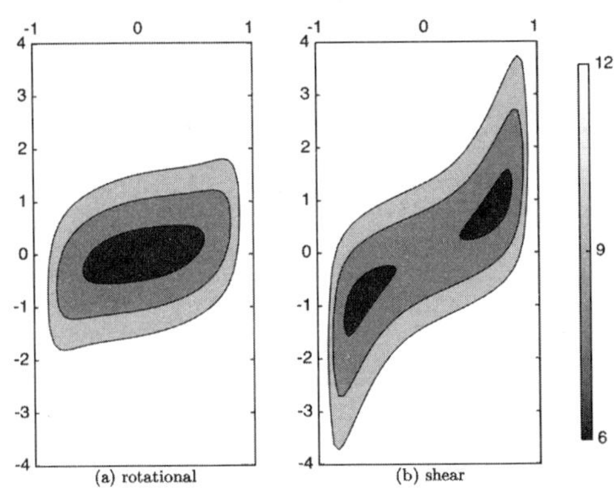

FIGURE 3.22. **Worm-Like Chain: Stationary Densities (Unknown).** These figures plot contours of the free energy of the numerical stationary density for a particle in the potential given by (3.7.9), and from left, rotational and shear flows given in (3.7.2) with $\gamma = 5$ and $\beta = 1$.

3.8. Adaptive Mesh Refinement in 2D

For the subsequent planar SDE problems with multiplicative noise, we use a variable step size grid with the property that the generator $\tilde{Q}_c$ in (2.4.7) evaluated on this grid is realizable. Recall, if the diffusion matrix is diagonally dominant, a grid with equal spacing in the horizontal and vertical directions suffices. If the diffusion matrix is constant, but not diagonally dominant, the ratio of the spacing in the horizontal and vertical directions can be chosen so that the grid has the desired property, since a 2×2, symmetric, and positive-definite matrix is always weakly diagonally dominant [**10**]. In general, if the diffusion matrix is neither diagonally

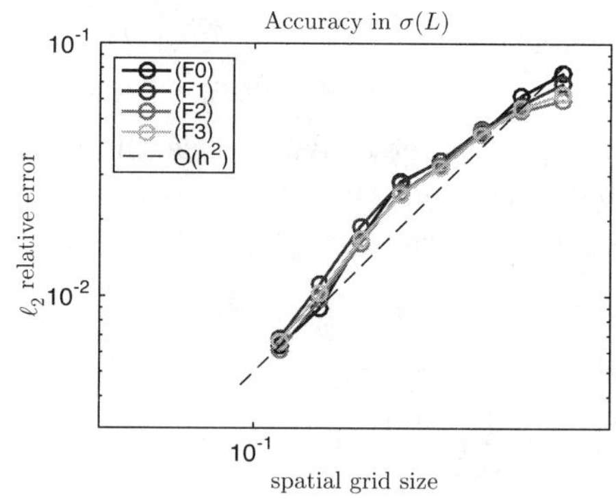

FIGURE 3.23. **Worm-Like Chain: Accuracy in $\sigma(L)$.** This figure plots the relative error in the ℓ_2 norm of the numerical estimate produced by $\tilde{Q}_c$ of the twenty eigenvalues of largest nontrivial real part of the operator L. Parameters are set at $\gamma = 1$ and $\beta = 1$. This figure suggests that $\tilde{Q}_c$ is about second-order accurate in representing the spectrum of the generator L.

dominant nor constant, it is not straightforward to construct a grid such that $\tilde{Q}_c$ is realizable on this grid. However, in the specific context of the log-normal and Lotka-Volterra examples considered in this paper, the following lemma states that a grid with the desired property exists. The key idea in this construction is to use a grid that is evenly spaced when viewed in logarithmic scaling.

To state this lemma, define:

$$\operatorname{diag}(x) = \begin{bmatrix} x_1 & 0 \\ 0 & x_2 \end{bmatrix} , \quad \text{for any } x = (x_1, x_2)^T \in \mathbb{R}^2.$$

LEMMA 3.8.1. *Consider the generator $\tilde{Q}_c$ in (2.4.7) with domain $\mathbb{R}^2_+$, an arbitrary drift field, and a diffusion matrix of the form:*

$$M(x) = \operatorname{diag}(x) \begin{bmatrix} M^{11} & M^{12} \\ M^{12} & M^{22} \end{bmatrix} \operatorname{diag}(x). \tag{3.8.1}$$

Assume that

$$M^{12} \neq 0 \,, \quad \det \begin{bmatrix} M^{11} & M^{12} \\ M^{12} & M^{22} \end{bmatrix} > 0 \,, \quad M^{11} + M^{22} > 0. \tag{3.8.2}$$

In this context, $\tilde{Q}_c$ in (2.4.7) is realizable on a gridded state space.

PROOF. To prove this lemma we construct a grid such that $\tilde{Q}_c$ in (2.4.7) is realizable on this grid. Let $S = \{(x_i, y_j)\} \subset \mathbb{R}^2_+$ be the desired grid. We define S to be the image of an evenly spaced grid in logspace:

$$\hat{S} = \{(\xi_i, \eta_j) \in \mathbb{R}^2 \mid \quad \delta\xi = \xi_{i+1} - \xi_i \,, \quad \delta\eta = \eta_{j+1} - \eta_j\}$$

where $\delta\xi$ and $\delta\eta$ set the grid size in the horizontal and vertical directions in logspace, respectively. We choose the spacing in logspace so that S has the desired property in physical space. Specifically, map every point $(\xi_i, \eta_j) \in \hat{S}$ to a point in $(x_i, y_j) \in S$

using the transformation: $x_i = \exp(\xi_i)$ and $y_j = \exp(\eta_j)$. Suppose that

$$\delta\xi = \alpha\epsilon \quad \text{and} \quad \delta\eta = \epsilon$$

where ϵ and α are positive parameters. We choose these parameters so that $\tilde{Q}_c$ evaluated on S is realizable. Note that the spacing between neighboring horizontal and vertical grid points in physical space satisfy:

$$(3.8.3) \qquad \begin{cases} \delta x_i^+ = x_{i+1} - x_i = (\exp(\alpha\epsilon) - 1)x_i \\ \delta x_i^- = x_i - x_{i-1} = (1 - \exp(-\alpha\epsilon))x_i \\ \delta x_i = \dfrac{\delta x_i^+ + \delta x_i^-}{2} = \sinh(\alpha\epsilon)x_i \end{cases}$$

and

$$(3.8.4) \qquad \begin{cases} \delta y_j^+ = y_{j+1} - y_j = (\exp(\epsilon) - 1)y_j \\ \delta y_j^- = y_j - y_{j-1} = (1 - \exp(-\epsilon))y_j \\ \delta y_j = \dfrac{\delta y_j^+ + \delta y_j^-}{2} = \sinh(\epsilon)y_j \end{cases}$$

Referring to (2.4.7), a necessary and sufficient condition for realizability of $\tilde{Q}_c$ is that all of the following inequalities hold:

$$(3.8.5) \qquad \begin{aligned} &\frac{M_{i,j}^{11}}{\delta x_i \delta x_i^+} - \frac{M_{i,j}^{12} \vee 0}{\delta x_i^+ \delta y_j^+} + \frac{M_{i,j}^{12} \wedge 0}{\delta x_i^+ \delta y_j^-} \geq 0 \\ &\frac{M_{i,j}^{11}}{\delta x_i \delta x_i^-} - \frac{M_{i,j}^{12} \vee 0}{\delta x_i^- \delta y_j^-} + \frac{M_{i,j}^{12} \wedge 0}{\delta x_i^- \delta y_j^+} \geq 0 \\ &\frac{M_{i,j}^{22}}{\delta y_j \delta y_j^+} - \frac{M_{i,j}^{12} \vee 0}{\delta x_i^+ \delta y_j^+} + \frac{M_{i,j}^{12} \wedge 0}{\delta x_i^- \delta y_j^+} \geq 0 \\ &\frac{M_{i,j}^{22}}{\delta y_j \delta y_j^-} - \frac{M_{i,j}^{12} \vee 0}{\delta x_i^- \delta y_j^-} + \frac{M_{i,j}^{12} \wedge 0}{\delta x_i^+ \delta y_j^-} \geq 0 \end{aligned}$$

Let us express these inequalities in terms of α and ϵ. Substitute (3.8.1) and the grid spacings (3.8.3) and (3.8.4) into (3.8.5) and simplify to obtain:

$$(3.8.6) \qquad \begin{aligned} &\frac{M^{11}}{\sinh(\alpha\epsilon)} - \frac{M^{12} \vee 0}{\exp(\epsilon) - 1} + \frac{M^{12} \wedge 0}{1 - \exp(-\epsilon)} \geq 0 \\ &\frac{M^{11}}{\sinh(\alpha\epsilon)} - \frac{M^{12} \vee 0}{1 - \exp(-\epsilon)} + \frac{M^{12} \wedge 0}{\exp(\epsilon) - 1} \geq 0 \\ &\frac{M^{22}}{\sinh(\epsilon)} - \frac{M^{12} \vee 0}{\exp(\alpha\epsilon) - 1} + \frac{M^{12} \wedge 0}{1 - \exp(-\alpha\epsilon)} \geq 0 \\ &\frac{M^{22}}{\sinh(\epsilon)} - \frac{M^{12} \vee 0}{1 - \exp(-\alpha\epsilon)} + \frac{M^{12} \wedge 0}{\exp(\alpha\epsilon) - 1} \geq 0 \end{aligned}$$

Applying the elementary inequality

$$\exp(a) - 1 > 1 - \exp(-a) \quad \forall\, a \in \mathbb{R}$$

to (3.8.6) yields the following sufficient condition for realizability

$$(3.8.7) \qquad \frac{M^{11}}{|M^{12}|} \geq \frac{\sinh(\alpha\epsilon)}{1 - \exp(-\epsilon)} \quad \text{and} \quad \frac{1 - \exp(-\alpha\epsilon)}{\sinh(\epsilon)} \geq \frac{|M^{12}|}{M^{22}}$$

The hypothesis in (3.8.2) implies that

$$M^{11} > 0\,, \quad M^{22} > 0 \quad \text{and} \quad \frac{M^{11}}{|M^{12}|} - \frac{|M^{12}|}{M^{22}} > 0.$$

Hence, we can choose α to satisfy:

$$\frac{M^{11}}{|M^{12}|} > \alpha > \frac{|M^{12}|}{M^{22}} > 0.$$

If ϵ is sufficiently small, these bounds on α imply that (3.8.7) holds. Indeed, a straightforward application of l'Hôpital's Monotonicity Rule implies that

$$\overline{\phi}(\epsilon) = \frac{\sinh(\alpha\epsilon)}{1 - \exp(-\epsilon)} \geq \alpha \geq \frac{1 - \exp(-\alpha\epsilon)}{\sinh(\epsilon)} = \underline{\phi}(\epsilon) \tag{3.8.8}$$

for any $\epsilon > 0$.

To prove the upper bound in (3.8.8), note that:

$$\lim_{s \to 0} \overline{\phi}(s) = \alpha.$$

Moreover, l'Hôpital's Monotonicity Rule implies that the function $\overline{\phi}(s)$ is increasing since the function:

$$s \mapsto \alpha \cosh(\alpha s) \exp(s)$$

is increasing with s. Similarly, $\underline{\phi}(s)$ is decreasing since the function:

$$s \mapsto \alpha \exp(-\alpha s) \operatorname{sech}(s)$$

is decreasing with s, and $\lim_{s \to 0} \underline{\phi}(s) = \alpha$. □

3.9. Log-normal Process in 2D with Multiplicative Noise

We derive a log-normal process by transforming a two-dimensional Ornstein-Uhlenbeck process. Thus, we obtain a log-normal process whose transition and equilibrium probability distributions are explicitly known and can be used as benchmarks to validate the approximation given by $\tilde{Q}_c$ in (2.4.7). First, let us recall some basic facts about Ornstein-Uhlenbeck processes. Consider a two-dimensional, non-symmetric Ornstein-Uhlenbeck process

$$dY = AY dt + \sqrt{2} B dW\,, \; Y(0) = x \in \mathbb{R}^2. \tag{3.9.1}$$

Let M be the diffusion matrix defined as:

$$M = BB^T = \begin{bmatrix} M^{11} & M^{12} \\ M^{12} & M^{22} \end{bmatrix}.$$

Recall that the transition probability of $Y(t)$ is a two-dimensional Gaussian distribution with mean vector $\xi(x,t)$ and covariance matrix $\Sigma(t)$ given by:

$$\xi(x,t) = \exp(tA)x\,,$$
$$\Sigma(t) = 2\int_0^t \exp(sA) M \exp(sA^T) ds$$

See, e.g., [**72**, Section 4.8]. To numerically compute $\Sigma(t)$ for any $t > 0$, observe that the covariance matrix $\Sigma(t)$ satisfies the Lyapunov equation:

$$\begin{aligned}\Sigma(t)A^T + A\Sigma(t) &= 2\left(\int_0^t \frac{d}{ds}\exp(sA)M\exp(sA^T)ds\right) ,\\ &= 2\left(\exp(tA)M(\exp(tA))^T - M\right) .\end{aligned}$$

The MATLAB command `lyap` can be used to solve equations of this type. Specifically, given the matrices $\mathcal{A}$ and $\mathcal{B}$ as inputs, `lyap` solves for the unknown matrix $\mathcal{X}$ which satisfies the equation:

$$\mathcal{A}\mathcal{X} + \mathcal{X}\mathcal{A}^T + \mathcal{B} = 0.$$

Defining the inputs to `lyap` as:

$$\mathcal{A} = A\ ,\ \ \mathcal{B} = -2\left(\exp(tA)M(\exp(tA))^T - M\right)\ ,$$

the function will output the desired covariance matrix. (A sufficient condition for a unique solution to this Lyapunov equation is that all of the eigenvalues of the matrix $\mathcal{A}$ have negative real parts.) If the eigenvalues of the matrix A in (3.9.1) all have strictly negative real part, passing to the limit in the above expressions as $t \to \infty$ implies that the equilibrium distribution is Gaussian with zero mean vector and a covariance matrix Σ_∞ that satisfies the Lyapunov equation:

$$\Sigma_\infty A^T + A\Sigma_\infty = -2M. \tag{3.9.2}$$

Consider the transformation of $Y(t) = (Y_1(t), Y_2(t))^T$ given by

$$X = (\exp(Y_1), \exp(Y_2))^T.$$

Under this transformation, the linear SDE (3.9.1) becomes:

$$dX = \mu(X)dt + \sqrt{2}\sigma(X)dW\ ,\ \ X(0) \in \mathbb{R}_+^2 \tag{3.9.3}$$

where the drift and diffusion fields are:

$$\begin{aligned}\mu(x) &= \begin{bmatrix} M^{11}x_1 \\ M^{22}x_2 \end{bmatrix} + \operatorname{diag}(x, y)A\begin{bmatrix} \log(x) \\ \log(x_2) \end{bmatrix}\\ M(x) &= (\sigma\sigma^T)(x)\\ &= \operatorname{diag}(x_1, x_2)\begin{bmatrix} M^{11} & M^{12} \\ M^{12} & M^{22} \end{bmatrix}\operatorname{diag}(x_1, x_2) = \begin{bmatrix} M^{11}x_1^2 & M^{12}x_1x_2 \\ M^{12}x_1x_2 & M^{22}x_2^2 \end{bmatrix}\end{aligned}$$

By change of variables, the (log-normal) probability and stationary densities of the transformed process $X(t)$ are given by:

$$\left.\begin{aligned}\tilde{p}_t(x) &= p_t(\log(x_1), \log(x_2))(x_1x_2)^{-1}\\ \tilde{\nu}(x) &= \nu(\log(x_1), \log(x_2))(x_1x_2)^{-1}\end{aligned}\right\} \qquad x = (x_1, x_2)^T \tag{3.9.4}$$

where $p_t(x)$ and $\nu(x)$ are the (normal) probability and stationary densities of the Ornstein-Uhlenbeck process $Y(t)$, respectively.

Figure 3.24 illustrates stationary density accuracy of $\tilde{Q}_c$ in (2.4.7) using the variable step size grid derived in the preceding Lemma 3.8.1.

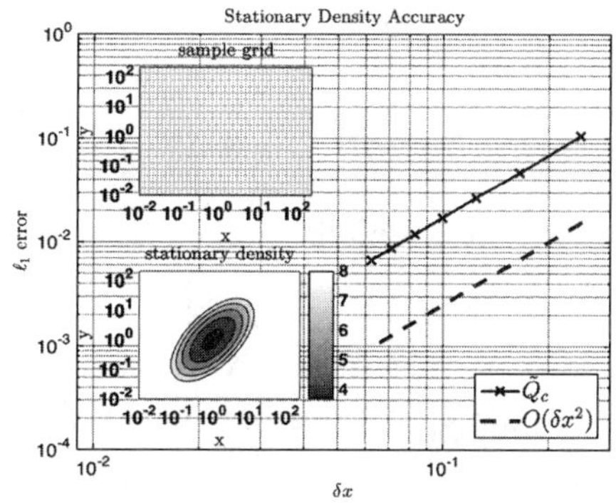

FIGURE 3.24. **Log-normal Process in 2D.** This figure illustrates stationary density accuracy of $\tilde{Q}_c$ in (2.4.7). The grid used is evenly spaced in logspace as shown in the upper left inset. Contours of the stationary density are plotted in the lower left inset. The spatial step size shown is with respect to the spacing of the grid in logspace. In this example the benchmark solution is the true stationary density given by the formula in (3.9.4). See Lemma 3.8.1 for more details about the grid.

3.10. Lotka-Volterra Process in 2D with Multiplicative Noise

Consider (2.1.2) with $\Omega = \mathbb{R}^2_+$ and set:

$$\mu(x) = \begin{bmatrix} k_1 x_1 - x_1 x_2 - \gamma_1 x_1^2 \\ -k_2 x_2 + x_1 x_2 - \gamma_2 x_2^2 \end{bmatrix}, \quad \text{and} \quad (\sigma\sigma^T)(x) = \operatorname{diag}(x) \begin{bmatrix} M^{11} & M^{12} \\ M^{12} & M^{22} \end{bmatrix} \operatorname{diag}(x).$$

This SDE is a model of interaction between a population of x_1 prey and x_2 predators that includes environmental fluctuations. Recall that when $\gamma_1 = \gamma_2 = 0$, the noise-free system reduces to an ODE with a fixed point at (k_2, k_1) and a first integral given by the following function: $V(x) = k_2 \log(x_1) + k_1 \log(x_2) - (x_1 + x_2)$. The diffusion field of this SDE satisfies the hypotheses in Lemma 3.8.1. Thus, we will be able to construct a grid with the property that $\tilde{Q}_c$ in (2.4.7) evaluated on this grid is realizable.

We use some known theoretical results on this SDE problem to benchmark the generator $\tilde{Q}_c$ [**139**]. For this purpose, it helps to define the parameters: $c_1 = k_1 - \frac{1}{2}M^{11}$ and $c_2 = k_2 + \frac{1}{2}M^{22}$. According to Theorem 1 of [**139**], for every $t > 0$ the probability distribution of the SDE is absolutely continuous with respect to Lebesgue measure on Ω. This property implies that the solution is strictly positive for all $t > 0$. This positivity, however, does not prevent the probability distribution of the solution from converging in the long time limit to a measure supported only on the boundaries of Ω, in which case one or both of the populations may become extinct. In fact, the finite-time probability distribution of the SDE solution converges to an invariant probability measure (IM) with the following support properties:

If $c_1 > \gamma_1 c_2$: then the IM is supported on $\mathbb{R}^2_+$.

If $\gamma_1 c_2 > c_1 > 0$: then the IM is supported on $\mathbb{R}_+ \times \{0\}$ with marginal density of x_1 given by:

$$f_\star(x_1) = Z^{-1} x_1^{\frac{2c_1}{M^{11}}-1} \exp\left(-\frac{2x_1}{M^{11}}\right) \tag{3.10.1}$$

where Z is the following normalization constant:

$$Z = \left(\frac{M^{11}}{2}\right)^{\frac{2c_1}{M^{11}}} \Gamma\left(\frac{2c_1}{M^{11}}\right) .$$

If $c_1 < 0$: then the IM is atomic at the origin.

Note that if $2c_1/M^{11} < 1$ (or $k_1 < M^{11}$), then the marginal stationary density in (3.10.1) is unbounded at $x = 0$. Figures 3.25 and 3.26 use $\tilde{Q}_c$ in (2.4.7) with the variable step size grid described in Lemma 3.8.1 to numerically verify these theoretical results.

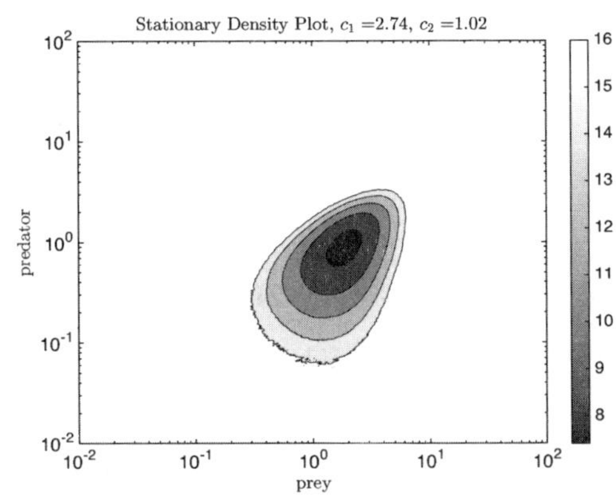

FIGURE 3.25. **Lotka-Volterra (LV) Process.** This figure plots contours of the numerical stationary density for the LV process produced by the generator $\tilde{Q}_c$ in (2.4.7) with the variable step size grid described in Lemma 3.8.1. The parameters are selected so that the true stationary distribution is supported on $\mathbb{R}^2_+$. This figure is in agreement with Theorem 1 of [**139**].

3.11. Colloidal Cluster in 39D with Multiplicative Noise

Consider a Brownian Dynamics (BD) simulation of a cluster consisting of 13 particles governed by the following SDE:

(3.11.1)

$$dY = -M(Y)D\mathcal{E}(Y)dt + \beta^{-1} \operatorname{div} M(Y)dt + \sqrt{2\beta^{-1}}\sigma(Y)dW \ , \ Y(0) \in \mathbb{R}^n \ ,$$
$$\sigma(x)\sigma(x)^T = M(x) \ ,$$

where $\mathcal{E}(x)$ is the potential energy of the system, $M(x)$ is the mobility matrix (or what we have been calling the diffusion matrix), and W is a n-dimensional Brownian motion. To be clear, the total dimension of this system is $n = 3 \times 13 = 39$. This BD model of a colloidal gel demonstrates that hydrodynamic interactions can influence the collapse dynamics of the cluster [**45**] as verified below.

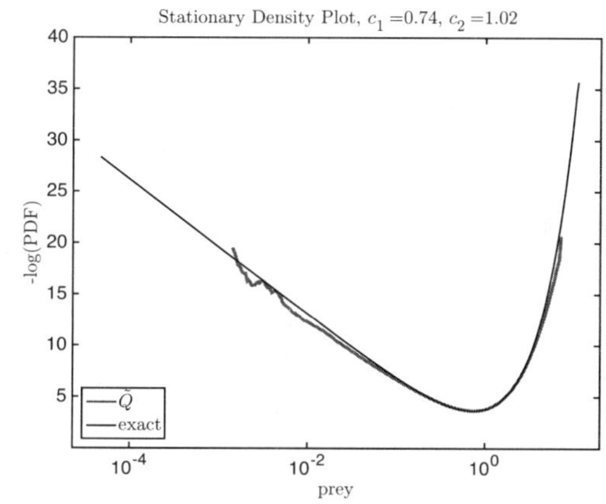

FIGURE 3.26. **Lotka-Volterra (LV) Process.** Here we numerically verify Theorem 1 of [**139**] using the generator $\tilde{Q}_c$ in (2.4.7) with the variable step size grid described in Lemma 3.8.1. The figure plots the density of the stationary distribution for the LV process in a parameter regime where it is supported on $\mathbb{R}_+ \times \{0\}$. In this case in the long time limit, the 'predator population' (or x_2 degree of freedom) becomes extinct, and an explicit formula for the stationary density in the 'prey population' (or x_1 degree of freedom) is given in (3.10.1), which we use to benchmark the numerically produced solution.

The potential energy $\mathcal{E}(x)$ involves short-range, pair-wise interactions between particles. It features a very steep hard core with a short-range attractive tail. When hydrodynamic interactions are accounted for, the mobility matrix depends on the state of the system and its entries involve long-range interactions that decay like $1/r$ with distance r between particle pairs. The precise form of $\mathcal{E}$ and M are derived below. In the situations we consider, the divergence of the mobility matrix appearing in (3.11.1) vanishes. Additionally, this diffusion is self-adjoint with respect to the density:

$$\nu(x) = \exp(-\beta\mathcal{E}(x)). \tag{3.11.2}$$

For background on the numerical treatment of self-adjoint diffusions with multiplicative noise see [**14**].

For this BD problem, numerical stability and accuracy impose demanding time step size requirements for time integrators due to the presence of singular coefficients in the drift of the governing SDE. Another issue with this SDE problem is that the hydrodynamic interactions between particles induce long-range interactions, which leads to an SDE with multiplicative noise.

Potential Energy. Consider a collection of N identical spherical particles with radius a. Let $q = (q_1, \dots, q_N) \in \mathbb{R}^{3N}$ denote the position of all N particles where $q_i \in \mathbb{R}^3$ for $i = 1, \dots, N$. The distance between the ith and jth particle is the usual Euclidean one:

$$d(q_i, q_j) = |q_i - q_j|. \tag{3.11.3}$$

In terms of this distance, the total energy is a sum of pair-wise potentials over all particle pairs:

$$\mathcal{E}(q) = \sum_{\substack{1 \le i \le n \\ j > i}} U(d(q_i, q_j)) \,, \tag{3.11.4}$$

Following the description in [**45**], we assume that the pair-wise potential energy $U(r)$ is given by:

$$U(r) = U_{\mathrm{SC}}(r) + U_{\mathrm{AO}}(r) \,, \tag{3.11.5}$$

which is a sum of a soft-core potential:

$$U_{\mathrm{SC}}(r) = \epsilon_{\mathrm{SC}} \left(\frac{2a}{r} \right)^{24} \,, \tag{3.11.6}$$

and an attractive potential that is defined in terms of the inter-particle force $F_{\mathrm{AO}}(r)$ it exerts as follows:

(3.11.7)

$$U_{\mathrm{AO}}(r) = \int_r^\infty F_{\mathrm{AO}}(s) ds \,, \quad F_{\mathrm{AO}}(r) = \begin{cases} c_{\mathrm{AO}}(D_2^2 - D_1^2) & \text{if } r < D_1 \\ c_{\mathrm{AO}}(D_2^2 - r^2) & \text{if } D_1 \le r < D_2 \\ 0 & \text{otherwise} \end{cases}$$

Here, we have introduced the parameters ϵ_{SC}, c_{AO}, D_1, and D_2. Figure 3.27 plots graphs of $U(r)$ and $F(r)$ for the parameter values given in Table 1. Note that $U(r)$ induces a short-range force. Further, because of the $1/r^{24}$-singularity in $U_{\mathrm{SC}}(r)$ in (3.11.6) at $r = 0$, this force makes the resulting SDE stiff. Since the energy is short-range, it is not uniformly coercive. Thus, the invariant density in (3.11.2) is not integrable on $\mathbb{R}^n$ and the system is not ergodic. In the numerical tests, we observe that – despite this non-ergodicity – for most sample paths, the cluster does not evaporate over the time-span of simulation.

Mobility Matrix. Consider again a collection of N identical spherical particles with radius a. Here we describe a commonly used, implicit model for the effect of the solvent known as the Rotne-Pragner-Yamakawa (RPY) approximation. This model results in solvent-mediated interactions between the N spherical particles through the mobility matrix $M(q)$. Let $I_{3\times 3}$ be the 3×3 identity matrix, R_{hydro} be the hydrodynamic radius of each bead, and η_s be the solvent viscosity. The RPY mobility matrix is given by:

$$M(q) = \begin{bmatrix} \Omega_{1,1} & \dots & \Omega_{1,N} \\ \vdots & \ddots & \vdots \\ \Omega_{N,1} & \dots & \Omega_{N,N} \end{bmatrix} \,, \quad \Omega_{i,j} = \begin{cases} \frac{1}{\zeta} I_{3\times 3} \,, & \text{if } i = j \\ \Omega_{RPY}(q_i - q_j) \,, & \text{otherwise} \end{cases} \tag{3.11.8}$$

for all $q \in \mathbb{R}^{3N}$. Here, we have introduced the 3×3 matrix $\Omega_{RPY}(x)$ defined as:

$$\Omega_{RPY}(x) = \frac{1}{\zeta} \left(C_1(x) I_{3\times 3} + C_2(x) \frac{x}{|x|} \otimes \frac{x}{|x|} \right) \tag{3.11.9}$$

where $C_1(x)$ and $C_2(x)$ are the following scalar-valued functions:

$$C_1(x) = \begin{cases} \frac{3}{4} \left(\frac{R_{\mathrm{hydro}}}{|x|} \right) + \frac{1}{2} \left(\frac{R_{\mathrm{hydro}}}{|x|} \right)^3 \,, & \text{if } |x| > 2R_{\mathrm{hydro}} \\ 1 - \frac{9}{32} \left(\frac{|x|}{R_{\mathrm{hydro}}} \right) \,, & \text{otherwise} \end{cases}$$

and,

$$C_2(x) = \begin{cases} \frac{3}{4}\left(\frac{R_{\text{hydro}}}{|x|}\right) - \frac{3}{2}\left(\frac{R_{\text{hydro}}}{|x|}\right)^3, & \text{if } |x| > 2R_{\text{hydro}} \\ \frac{3}{32}\left(\frac{|x|}{R_{\text{hydro}}}\right), & \text{otherwise} \end{cases}$$

The quantity $1/\zeta$ is the mobility constant produced by a single bead translating in an unbounded solvent at a constant velocity: $\zeta = 6\pi\eta_s R_{\text{hydro}}$. The approximation (3.11.8) preserves the physical property that the mobility matrix is positive definite, satisfies $(\operatorname{div} M)(q) = 0$ for all $q \in \mathbb{R}^{3N}$, and is exact up to $\mathcal{O}((R_{\text{hydro}}/r_{ij})^4)$ where r_{ij} is the distance between distinct particles i and j. To read more about the RPY approximation see [**138**].

Description of Numerical Test. We are now in position to repeat the simulation in [**32**, **45**] using the generator Q_c in (2.5.2). Consider 13 particles initially placed on the vertices and center of an icosahedron of edge length 8.08, which corresponds to a radius of gyration of 7.68. Note that this edge length is slightly greater than the equilibrium bond length, which is 6.4. Thus, without noise, the cluster would simply relax to an icosahedron with an edge length equal to the equilibrium bond length. To prevent exploding trajectories that can occur if any two particles get too close, we set the spatial grid size to be 10% of the equilibrium radius or 0.32. This choice ensures that the numerical approximation to the repulsive force cannot cause any particle to take a single spatial step that is larger than the cutoff radius of the short-range interaction. Thus, the approximation to the repulsive force between every particle pair is sufficiently resolved in space to prevent explosion. Figure 3.28 and 3.29 illustrate the results produced by an SSA integrator induced by the generator Q_c in (2.5.2) operated at this spatial step size. Other simulation parameters are given in Table 1. An important physical parameter is the Brownian time-scale defined as: $t_B = \beta\eta_s a^3$. In order to compare to the data from [**32**], the time span of simulation is long and set equal to $T = 400 t_B = 1065$. We emphasize that the spatial grid size with and without hydrodynamic interactions is set equal to $10\% a$ or $\delta x = 0.32$. Finally, Figure 3.30 validates Proposition 4.3.11, which quantifies the average number of computational steps of the SSA integrator as a function of the spatial step size.

Parameter	Description	Value(s)
	Physical Parameters	
N	# of particles	13
η_s	solvent viscosity	1
β^{-1}	temperature factor	12.3
a	**particle radius**	3.2
R_{hydro}	hydrodynamic radius	$\{a, 0\}$
ϵ_{SC}	parameter in U_{SC}	10
D_1	parameter in U_{AO}	$2.245a$
D_2	parameter in U_{AO}	$2.694a$
c_{AO}	parameter in U_{AO}	$58.5/a^3$
	Numerical Parameters	
δx	**spatial step size**	10% a
T	**time-span of simulation**	1065

TABLE 1. **Colloidal Collapse Simulation Parameters.** Physical parameter values are taken from: [**32**, **45**].

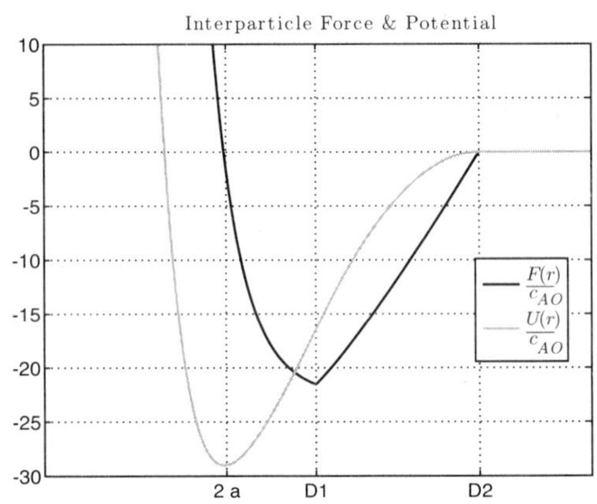

FIGURE 3.27. **Graph of pair-wise potential energy and force.** Graph of the inter-particle energy $U(r)/c_{\text{AO}}$ given in (3.11.5) and corresponding inter-particle force $F(x)/c_{\text{AO}} = -U'(r)/c_{\text{AO}}$ as a function of the inter-particle distance r with $c_{\text{AO}} = 1.7853$ and $a = 3.2$. The range of interaction is specified by the parameter D_2, which here is set equal to $D_2 = 2.694a \approx 8.6$. The energy is minimized at twice the particle radius a with minimum value: $U(2a) \approx -51.82$. Up to a factor c_{AO}, this graph agrees with Fig. S1 in the Supplementary Material accompanying [**45**].

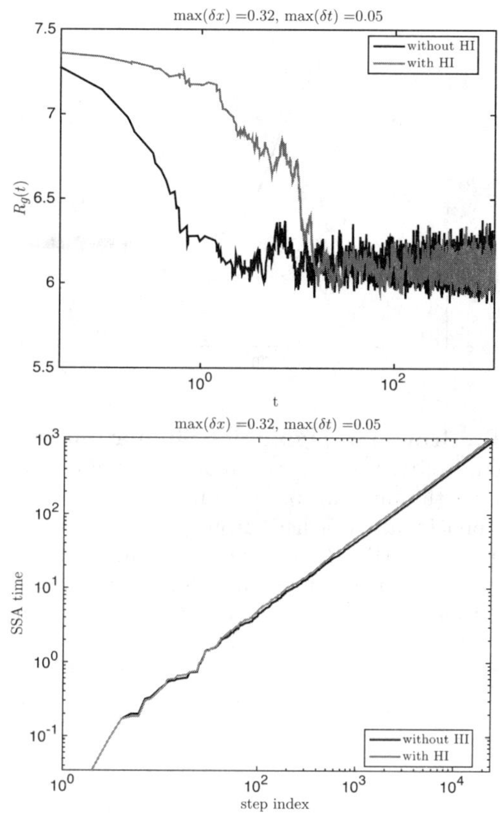

FIGURE 3.28. **Brownian Dynamics Simulation in 39D.** This figure shows sample paths generated by an SSA integrator of the collapse of a colloidal cluster. The top panel shows a sample trajectory with and without hydrodynamic interactions (HI). The bottom panel shows the SSA time vs step index for these sample trajectories. These sample paths are produced using the SSA induced by the generator Q_c in (2.5.2) operated at a spatial step size set equal to $10\%a$, where $2a$ is the equilibrium bond length from Figure 3.27.

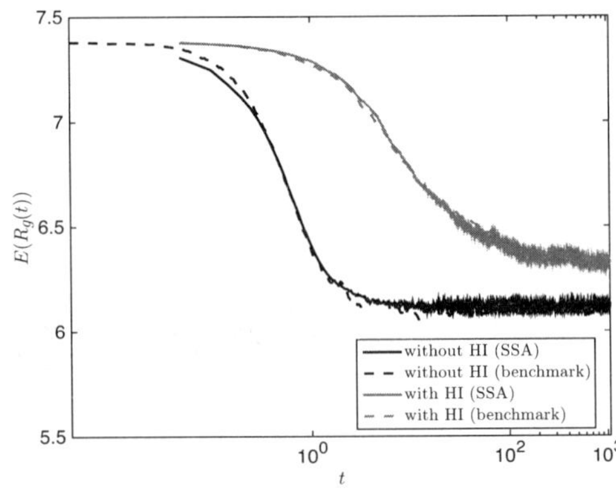

FIGURE 3.29. **Brownian Dynamics Simulation in 39D.** This figure shows the mean radius of gyration of the colloidal cluster as a function of time over the interval $[0, T]$ computed using an SSA integrator (solid) and a benchmark (dashed) from [**45**] with and without hydrodynamic interactions (HI). This numerical result is produced using the SSA induced by the generator Q_c in (2.5.2) operated at a spatial step size set equal to $10\%a$, where $2a$ is the equilibrium bond length from Figure 3.27.

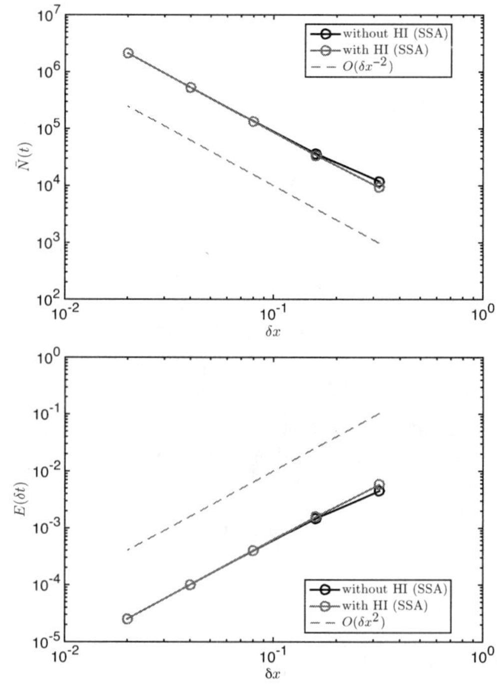

FIGURE 3.30. **Complexity of SSA in Brownian Dynamics Simulation in 39D.** The top panel of this figure plots the mean number of computational steps of the SSA integrator as a function of the spatial step size. The bottom panel of this figure plots the mean holding time of the SSA integrator as a function of the spatial step size. The figure validates Proposition 4.3.11. This numerical result is produced using the SSA induced by the generator Q_c in (2.5.2).

CHAPTER 4

Analysis on Gridded State Spaces

In this chapter we analyze the stability, complexity, and accuracy of the realizable discretization Q_c in (2.5.2) on a gridded state space. (Refer to §2.2 for the definition of a gridded state space.) For the stability analysis, we assume that the drift field is locally Lipschitz continuous and weakly dissipative. To show that the approximation is second-order accurate, we further assume that the drift field is of class C^4. In order to ensure that the state space is gridded, we assume that the noise field is constant. These assumptions are sufficient to prove that the approximation has a stochastic Lyapunov function, which is then used to quantify the complexity and accuracy of the approximation.

4.1. Assumptions

Here we describe the main objects of the theory in this chapter:

- generators of the processes, given in (4.1.4) and (4.1.8);
- stochastic equations governing the dynamics of sample paths, given in (4.1.5) and (4.1.12); and,
- Kolmogorov equations governing the evolution of conditional expectations of an observable, given in (4.1.6) and (4.1.15).

4.1.1. Some Notation. We shall use the following notation. Let $|\cdot|$ denote the 2-norm of an n-vector, let $|\cdot|_p$ denote the p-norm of an n-vector where $p \geq 1$ and let $\|\cdot\|$ denote the standard operator norm corresponding to $|\cdot|$. For a real-valued function $f \in C^r(\mathbb{R}^n)$ where $r \in \mathbb{N}$, let $D^r f$ denote the rth derivative of f. Let $B_b(\mathbb{R}^n)$ denote the Banach space of all Borel measurable and bounded functions from $\mathbb{R}^n$ to $\mathbb{R}$ endowed with the supremum norm:

$$\|f\|_\infty = \sup_{x \in \mathbb{R}^n} |f(x)| \,, \quad f \in B_b(\mathbb{R}^n). \tag{4.1.1}$$

We denote by $C_b(\mathbb{R}^n)$ the subspace of $B_b(\mathbb{R}^n)$ consisting of all uniformly continuous and bounded functions. This subspace with the supremum norm:

$$\|f\|_\infty = \sup_{x \in \mathbb{R}^n} |f(x)| \,, \quad f \in C_b(\mathbb{R}^n) \,, \tag{4.1.2}$$

is a Banach space. For $k \in \mathbb{N}$, we denote by $C_b^k(\mathbb{R}^n)$ the subspace of all k-times differentiable functions whose derivatives up to kth order are uniformly continuous and bounded in the supremum norm. This subspace is also a Banach space when endowed with the norm:

$$\|f\|_k = \|f\|_\infty + \sum_{i=1}^{k} \sup_{x \in \mathbb{R}^n} \|D^i f(x)\| \,, \quad f \in C_b^k(\mathbb{R}^n). \tag{4.1.3}$$

4.1.2. SDE. Consider a diffusion operator L, which maps a twice differentiable test function $f : \mathbb{R}^n \to \mathbb{R}$ to another function $Lf : \mathbb{R}^n \to \mathbb{R}$ defined as:

$$\boxed{Lf(x) = \operatorname{trace}(Df(x)\mu(x)^T + D^2 f(x))} \tag{4.1.4}$$

where $\mu : \mathbb{R}^n \to \mathbb{R}^n$ is the drift field. The SDE associated to L has an unbounded domain $\mathbb{R}^n$, isotropic (or direction-independent) and additive (or state-independent) noise, and can be written as:

$$\boxed{dY = \mu(Y)dt + \sqrt{2}\, dW \ , \quad Y(0) = x \in \mathbb{R}^n} \tag{4.1.5}$$

Under certain assumptions on the drift field (e.g. those stated in Assumption 4.1), this SDE possesses a strong-sense solution that is strong-sense unique. Basic properties of solutions to (4.1.5) are reviewed in §4.3.

Define the domain of L to be the following subset of measurable functions:

$$\operatorname{Dom}(L) =$$
$$\left\{ f \in B(\mathbb{R}^n) \ \middle| \ \left(f(Y(\cdot)) - \int_0^{\cdot} Lf(Y(s))ds \right) \begin{array}{c} \text{is a martingale for every} \\ \text{initial condition } Y(0) \in \mathbb{R}^n \end{array} \right\} .$$

For any $t \geq 0$ and $x \in \mathbb{R}^n$, let P_t and $\Pi_{t,x}$ resp. denote the Markov semigroup and Markov transition probabilities of $Y(t)$ related by:

$$P_t f(x) = \int_{\mathbb{R}^n} f(y)\Pi_{t,x}(dy) = \mathbb{E}_x f(Y(t)) \quad t \geq 0 \ , \ x \in \mathbb{R}^n \ , \ f \in \operatorname{Dom}(L).$$

Here $\mathbb{E}_x$ denotes expectation conditional on $Y(0) = x \in \mathbb{R}^n$.

The Kolmogorov equation associated to the SDE problem (4.1.5) is given by:

$$\boxed{\frac{\partial u}{\partial t}(x,t) = Lu(x,t) \quad \text{for all } x \in \mathbb{R}^n \text{ and } t \geq 0} \tag{4.1.6}$$

with initial condition:

$$u(x,0) = f(x) \quad \text{for all } x \in \mathbb{R}^n \text{ and for any } f \in \operatorname{Dom}(L).$$

The solution $u(x,t)$ to (4.1.6) has the following stochastic representation:

$$u(x,t) = \mathbb{E}_x f(Y(t)) = P_t f(x).$$

Under suitable conditions on the drift field (e.g. those stated in Assumption 4.1), the process $Y(t)$ admits a unique invariant probability measure Π that is absolutely continuous with respect to Lebesgue measure on $\mathbb{R}^n$. Let $\nu(x)$ be the stationary density associated to Π that satisfies

$$(L^*\nu)(x) = 0 \quad \text{for all } x \in \mathbb{R}^n$$

where L^* is the (formal) adjoint of the infinitesimal generator L of the SDE solution. Hence,

$$\Pi(Lf) = \int_{\mathbb{R}^n} Lf(x)\nu(x)dx = 0 \ , \quad \text{for all } f \in \operatorname{Dom}(L).$$

For the associated semigroup this invariance implies:

$$\Pi(P_t f) = \int_{\mathbb{R}^n} P_t f(x)\nu(x)dx = \int_{\mathbb{R}^n} f(x)\nu(x)dx = \Pi(f)$$

for any $t \geq 0$.

Next we introduce (space) discrete analogs of (4.1.4), (4.1.5), and (4.1.6).

4.1.3. Approximation. To introduce the generator of the approximation, let $\{e_i\}_{i=1}^n$ be the standard basis on $\mathbb{R}^n$ and let S_h be an infinite Cartesian grid over $\mathbb{R}^n$ with edge length h, which we refer to as the spatial step size. The precise location of this grid is typically set by the choice of initial condition. Let $\ell^\infty(S_h)$ (or ℓ^∞ for short) be the Banach space of all grid functions $\varphi : S_h \to \mathbb{R}$ that are bounded in the supremum norm:

$$\|\varphi\|_\infty = \sup_{x \in S_h} |\varphi(x)|. \tag{4.1.7}$$

In the context of the SDE problem (4.1.5), the infinitesimal generator of the approximation given in (2.5.2) maps a grid function $\varphi : S_h \to \mathbb{R}$ to another grid function $Q\varphi : S_h \to \mathbb{R}$ defined as:

$$\boxed{\begin{aligned} Q\varphi(x) = \sum_{i=1}^n & \frac{1}{h^2} \exp\left(\frac{h}{2}\mu_i\right) (\varphi(x + he_i) - \varphi(x)) \\ &+ \frac{1}{h^2} \exp\left(-\frac{h}{2}\mu_i\right) (\varphi(x - he_i) - \varphi(x)) \end{aligned}} \tag{4.1.8}$$

Here we used the shorthand notation $\mu_i = \mu(x)^T e_i$. We also dropped the subscript c appearing in (2.5.2), since the theory in this chapter focuses on proving properties of this generator only. For any $t \geq 0$, the generator Q induces a Markov jump process $X(t)$ on S_h, which can be exactly simulated using the SSA specified in Algorithm 2.1. For basic facts about Markov jump processes, we refer the reader to [**41**, **83**, **131**, **133**].

Since our assumptions (given below) permit the drift field entering the jump rates in (4.1.8) to be unbounded, Q may not be a bounded operator with respect to grid functions in ℓ^∞. Nevertheless under a weak dissipativity condition on the drift field of (4.1.5) given in Assumption 4.1, we show that the approximation $X(t)$ has a stochastic Lyapunov function. This property is then used to show that the local martingale

$$M^\varphi(t) = \varphi(X(t)) - \varphi(X(0)) - \int_0^t Q\varphi(X(s))ds \,, \quad t \geq 0 \tag{4.1.9}$$

is a global martingale for all $\varphi : S_h \to \mathbb{R}$ satisfying a growth condition, which is developed in §4.3. For functions satisfying this condition, Dynkin's formula holds:

$$\mathbb{E}_x \varphi(X(t)) = \varphi(x) + \int_0^t \mathbb{E}_x Q\varphi(X(s))ds \,, \quad t \geq 0. \tag{4.1.10}$$

This class of functions includes locally Lipschitz continuous functions that have a local Lipschitz constant that does not grow faster than a sharp stochastic Lyapunov function for the approximation.

Note that the compensator of the quadratic variation of M^φ is given by

$$\langle M^\varphi, M^\varphi \rangle(t) = \int_0^t \Big(Q\varphi^2(X(s)) - 2\varphi(X(s))Q\varphi(X(s)) \Big) \, ds. \tag{4.1.11}$$

By choosing $\varphi(X(t))$ in (4.1.9) to be the ith-component of $X(t)$, it can be shown that the process $X(t)$ induced by the generator Q satisfies the following stochastic equation:

$$\boxed{dX = \mu^h(X)dt + \sqrt{2}\, d\mathcal{M}\,, \quad X(0) = x \in S_h} \tag{4.1.12}$$

where we have introduced the drift field μ^h with ith component

$$\mu_i^h(x) = \frac{2}{h} \sinh\left(\frac{h}{2}\mu_i(x)\right) \quad \text{for } 1 \le i \le n \tag{4.1.13}$$

and a local (zero-mean) martingale

$$\mathcal{M}(t) = (\mathcal{M}_1(t), \ldots, \mathcal{M}_n(t))^T \in \mathbb{R}^n.$$

The compensator of the quadratic covariation of the components of $\mathcal{M}$ satisfies:

$$\langle \mathcal{M}_i, \mathcal{M}_j \rangle(t) = \begin{cases} 2\displaystyle\int_0^t \cosh\left(\frac{h}{2}\mu_i(X(s))\right) ds & \text{if } i = j \\ 0 & \text{otherwise} \end{cases} \tag{4.1.14}$$

for all $t \ge 0$. (Since jumps only occur in each coordinate direction, the compensator of the quadratic covariation of two distinct components of $\mathcal{M}$ is zero.) We stress that (4.1.12) is shorthand notation for a semimartingale representation of the Markov jump process $X(t)$. The noise entering (4.1.12) is purely discontinuous and the drift field is less regular than in the original SDE. However, this stochastic equation has the nice feature that its solutions can be computed (in law) by the SSA procedure specified in Algorithm 2.1. This stochastic equation resembles an ODE with random impulses [**140**]. Properties of strong solutions to (4.1.12) are analyzed in §4.3. We also remark that although there is a $1/h$ prefactor in μ^h, to leading order $\mu^h \approx \mu$, as expected.

Note that Q evaluated on the grid S_h can be represented as an infinite matrix Q with the following nonzero off-diagonal transition rates from state $x \in S_h$ to state $y \in S_h$:

$$\mathsf{Q}(x, y) = \frac{1}{h^2} \exp\left(\frac{1}{2}\mu(x)^T(y - x)\right) \quad \text{if } |y - x| = h$$

or in words, if y is a nearest neighbor of x. The Q-matrix property given in (2.2.1) implies that:

$$\mathsf{Q}(x, x) = -\sum_{y \in S_h \setminus \{x\}} \mathsf{Q}(x, y).$$

Recall from Algorithm 2.1, that the diagonal term of Q is the reciprocal of the mean holding time of the approximation $X(t)$ at the state x.

Consider the semi-discrete (discrete in space and continuous in time) Kolmogorov equation:

$$\boxed{\frac{du^h}{dt}(x, t) = Qu^h(x, t) \quad \text{for all } x \in S_h \text{ and } t \ge 0} \tag{4.1.15}$$

with initial condition:

$$u^h(x, 0) = \varphi(x) \quad \text{for all } x \in S_h.$$

Analogous to the continuous case, this solution can be explicitly represented in terms of $X(t)$:

$$u^h(x, t) = \mathbb{E}_x \varphi(X(t)) = P_t^h \varphi(x)$$

where we have introduced P_t^h, which is the Markov semigroup associated to Q. Define the domain of Q as the following subset of grid functions:

$$\mathrm{Dom}(Q) =$$

$$\left\{ \varphi : S_h \to \mathbb{R} \;\middle|\; \left(\varphi(X(\cdot)) - \int_0^{\cdot} Q\varphi(X(s))ds \right) \begin{array}{c}\text{is a martingale for every} \\ \text{initial condition } X(0) \in \mathbb{R}^n\end{array} \right\}.$$

Let Π^h denote an invariant probability measure of Q, which we will prove exists under certain conditions on the drift field stated below in Assumption 4.1. In the particular case of a gridded state space, Π^h is also unique. In contrast, if the state space is gridless, uniqueness is nontrivial to prove because the process may lack irreducibility. We address this point in Chapter 5.

Let $\nu^h(x)$ be the probability density function associated to Π^h, which satisfies the identity

$$(Q^* \nu^h)(x) = 0 \quad \text{for all } x \in S_h$$

where Q^* is the adjoint of Q. This identity implies that

$$\Pi^h(Q\varphi) = 0 \qquad \text{for all } \varphi \in \mathrm{Dom}(Q).$$

Moreover, an invariant measure Π^h satisfies:

$$\Pi^h(P_t^h \varphi) = \sum_{x \in S_h} P_t^h \varphi(x) \nu^h(x) = \sum_{x \in S_h} \varphi(x) \nu^h(x) = \Pi^h(\varphi)$$

for any $t \geq 0$.

Table 1 summarizes our notation.

TABLE 1. **Tabular summary of (space) continuous & discrete objects.**

	process	generator	transition probability	semigroup	invariant measure & density
SDE Solution	$Y(t) \in \mathbb{R}^n$	L	$\Pi_{t,x}$	P_t	$\Pi(A) = \int_A \nu(x)dx$ $A \subset \mathbb{R}^n$
Numerical Approximation	$X(t) \in S_h$	Q	$\Pi_{t,x}^h$	P_t^h	$\Pi^h(A) = \sum_{x \in A} \nu^h(x)$ $A \subset S_h$

4.1.4. Assumptions on Drift Field. Here we state and illustrate our assumptions on the drift field. In this assumption, note that we write a vector $x \in \mathbb{R}^n$ in components as $x = (x_1, \cdots, x_n)^T$.

ASSUMPTION 4.1. The drift field $\mu : \mathbb{R}^n \to \mathbb{R}^n$ of the SDE in (4.1.5) is of class C^4 and satisfies conditions (A1) - (A4) given below.

(A1) One-Sided Lipschitz Continuity: For all $x, \xi \in \mathbb{R}^n$, there exists $C_L^\mu \in \mathbb{R}$ such that:

$$D\mu(x)(\xi, \xi) \leq C_L^\mu |\xi|^2.$$

(A2) Maximal Growth: For any $1 \leq k \leq 4$, there exist $m \in \mathbb{N}_0$ and $C_P^\mu > 0$ such that:

$$\sup_{x \in \mathbb{R}^n} \frac{|\mu(x)|}{1 + |x|^{2m+1}} \vee \sup_{x \in \mathbb{R}^n} \frac{\|D^k \mu(x)\|}{1 + |x|^{2m+1-k}} \leq C_P^\mu.$$

(A3) Weak Dissipativity: For all $x \in \mathbb{R}^n$ and for every $1 \le i \le n$, there exist $\beta_0 > 0$ and $\alpha_0 \ge 0$ such that:

$$\mu_i(x)\,\mathrm{sign}(x_i) \le \alpha_0\,(1+|x|^{2m}) - \beta_0\,|x_i|^{2m+1} \tag{4.1.16}$$

where m is the natural number appearing in (A2).

(A4) Minimal Growth: For all $x \in \mathbb{R}^n$ and for every $1 \le i \le n$,

$$|\mu_i(x)| \ge -\alpha_0\,(1+|x|^{2m}) + \beta_0\,|x_i|^{(2m+1)}$$

where m is the natural number appearing in (A2) and α_0, β_0 are the constants appearing in (A3).

We chose this assumption with the following drift fields in mind.

EXAMPLE 1. Let $g : \mathbb{R}^n \to \mathbb{R}^n$ be of class C^4 with component form

$$g(x) = (g_1(x), \ldots, g_n(x))^T$$

that satisfies the maximal polynomial growth condition:

$$\sup_{x\in\mathbb{R}^n} \frac{|g(x)|}{1+|x|^{2m}} \vee \sup_{x\in\mathbb{R}^n} \frac{\|D^k g(x)\|}{1+|x|^{2m-k}} < C_P^g \quad \text{for any } 1 \le k \le 4 \tag{4.1.17}$$

and the one-sided Lipschitz property:

$$Dg(x)(\xi,\xi) \le C_L^g |\xi|^2 \quad \text{for all } x, \xi \in \mathbb{R}^n \tag{4.1.18}$$

for some $m \in \mathbb{N}_0$ and for some constants $C_P^g > 0$ and $C_L^g \in \mathbb{R}$.

The drift field with ith-component given by:

$$\boxed{\mu_i(x) = -a_i\, x_i^{2m+1} + g_i(x)\,, \quad a_i > 0\,, \quad 1 \le i \le n}$$

satisfies Assumption 4.1. Indeed,

$$\begin{aligned} D\mu(x)(\xi,\xi) &= \sum_{1\le i,j\le n} \frac{\partial \mu_i}{\partial x_j} \xi_i \xi_j \\ &= \sum_{1\le i,j\le n} -a_i\,(2m+1)\,x_i^{2m}\,\xi_i^2\,\delta_{ij} + \frac{\partial g_i}{\partial x_j}\xi_i\xi_j \le C_L^g |\xi|^2 \end{aligned}$$

where δ_{ij} is the Kronecker delta and in the last line we used (4.1.18). This inequality implies that Assumption 4.1 (A1) holds with $C_L^\mu = C_L^g$. Moreover, according to our hypotheses on g, it is apparent that Assumption 4.1 (A2) holds. We also have that

$$\begin{aligned} \mu_i(x)\,\mathrm{sign}(x_i) &= -a_i\,|x_i|^{2m+1} + g_i(x)\,\mathrm{sign}(x_i) \\ &\le -a_i\,|x_i|^{2m+1} + C_P^g\,(1+|x|^{2m}) \end{aligned}$$

for all $x \in \mathbb{R}^n$, which implies that Assumption 4.1 (A3) holds with $\alpha_0 = C_P^g$ and $\beta_0 = \min_{1\le i\le n} a_i$. By the reverse triangle inequality, we have

$$\begin{aligned} |\mu_i(x)| &\ge -|g_i(x)| + a_i\,|x_i|^{2m+1} \\ &\ge -\alpha_0\,(1+|x|^{2m}) + \beta_0\,|x_i|^{2m+1} \end{aligned}$$

which implies that Assumption 4.1 (A4) holds.

If μ is differentiable, then Assumption 4.1 (A1) is equivalent to the following condition.

LEMMA 4.1.1. *Suppose that μ is C^1. Then Assumption 4.1 (A1) implies that μ satisfies*

$$\langle \mu(y) - \mu(x), y - x \rangle \leq C_L^{\mu} |x - y|^2 \quad \textit{for all } x, y \in \mathbb{R}^n$$

where C_L^{μ} is the constant appearing in Assumption 4.1 (A1).

PROOF. Use Taylor's formula and Assumption 4.1 (A1) to obtain:

$$\begin{aligned}\langle \mu(y) - \mu(x), y - x \rangle &= \int_0^1 D\mu(x + s(y - x))(y - x, y - x) ds \\ &\leq C_L^{\mu} \, |y - x|^2\end{aligned}$$

as required. □

REMARK 4.1.1 (On Usage of Assumption 4.1). Local Lipschitz continuity of the drift field is sufficient to show that for any initial condition $x \in \mathbb{R}^n$ and for any Brownian motion W there a.s. exists a unique process Y adapted to W that satisfies (4.1.5) up to a random explosion time. The weak dissipativity condition (A3) is used in Lemma 4.2.3 to prove that Y possesses a stochastic Lyapunov function, which implies that this pathwise unique solution can be extended to any $t \geq 0$ [**78**]. Additionally, Lemma 4.2.3 is used in Lemma 4.3.1 to obtain bounds on every moment of the solution that are uniform in time.

Assumption (A1) and (A2) are used to prove Lemma 4.3.6, which implies that the semigroup P_t is strongly Feller. Local Lipschitz continuity of the drift field is also used in Lemma 4.2.4 to prove that the process is irreducible and its transition probabilities are absolutely continuous with respect to Lebesgue measure. The strong Feller property and irreducibility imply uniqueness of an invariant probability measure. In fact, a stochastic Lyapunov function, irreducibility, and the strong Feller property imply that the process Y is geometrically ergodic with respect to this unique invariant probability measure, as stated in Theorem 4.2.5.

Regarding the stability of the approximation, Assumptions (A2), (A3) and (A4) are used in Lemmas 4.2.6 and 5.3.1 to prove that the approximation has a stochastic Lyapunov function. In the particular case of gridded state space, the approximation is also irreducible, and it then follows that the approximation is geometrically ergodic with respect to a unique invariant probability measure as stated in Theorem 4.2.8. Assumption (A2) is used to bound the growth of μ by this stochastic Lyapunov function in Lemmas 4.3.8 and 4.3.11. In the gridless state space context, Assumption (A2) is also used in Lemma 5.5.1 to show that a mollified generator is a bounded linear operator in the standard operator norm. This property implies that the semigroup of the approximation is Feller, and together with a stochastic Lyapunov function, that the approximation has an invariant probability measure.

Regarding the accuracy of the approximation, we require the additional assumption that the drift field has four derivatives. This assumption is used in Lemmas 4.3.5 and 4.3.6. The latter Lemma implies that if $f \in C_b^4(\mathbb{R}^n)$ then $P_t f \in C_b^4(\mathbb{R}^n)$ for any $t \geq 0$. This result is then used to estimate the global error of the approximation over finite (resp. infinite) time intervals in Theorem 4.5.1 (resp. Theorem 4.5.2). Four derivatives of the drift field are needed for proving second-order accuracy because the remainder term in $(Qf(x) - Lf(x))$ involves four derivatives of $f(x)$, see Lemma 4.4.1.

The growth conditions appearing in this assumption are convenient to work with. Relaxing these conditions would require retracing the steps taken in [**23**] to prove estimates on the regularity of the solution (which we rely on) and replacing every usage of this growth condition with a weak dissipativity-type condition.

The following Lemma relates our weak dissipativity assumption to more standard ones. Note that the converse implication does not hold, in general.

LEMMA 4.1.2. *Assumption 4.1 (A3) implies that there exist constants $\alpha \geq 0$ and $\beta > 0$ such that*

$$\mu(x)^T x \leq \alpha\,(1 + |x|^{2m+1}) - \beta\,|x|^{2m+2} \quad \textit{for all } x \in \mathbb{R}^n$$

where m is the natural number appearing in Assumption 4.1 (A2).

PROOF. Multiply both sides of (4.1.16) by $|x_i|$ and sum over i to obtain:

$$\mu(x)^T x \leq \alpha_0\,(1 + |x|^{2m})|x|_1 - \beta_0\,|x|_{2m+2}^{2m+2}.$$

Since norms on $\mathbb{R}^n$ are equivalent, there exist $C_1, C_2 > 0$ such that:

$$|x|_1 \leq C_1|x| \quad \text{and} \quad |x|_{2m+2}^{2m+2} \geq C_2|x|^{2m+2} \quad \text{for all } x \in \mathbb{R}^n.$$

These inequalities combined with the preceding one imply the desired inequality holds with $\alpha = 2C_1\alpha_0$ and $\beta = C_2\beta_0$. □

4.2. Stability by Stochastic Lyapunov Function

In this section we show that the approximation and the SDE solution are long-time stable under Assumption 4.1 on the drift field. Specifically, we show that the Markov processes induced by the generators L in (4.1.4) and Q in (4.1.8) are geometrically ergodic. The model problem considered for this analysis is (4.1.5), which is an SDE of elliptic type with an unbounded state space, locally Lipschitz drift coefficients, and additive noise. The weak dissipativity condition Assumption 4.1 (A3) is a crucial ingredient to prove that both the SDE solution and numerical approximation are geometrically ergodic. We emphasize that for this model problem the state space of the approximation is gridded and the generator Q can be represented by an infinite matrix. Our main tool to prove geometric ergodicity is Harris Theorem, which we recall below. We use the total variation (TV) metric to quantify the distance between probability measures.

DEFINITION 4.2.1. If Π_1 and Π_2 are two probability measures on a measurable space Ω, their total variation is defined as:

$$\|\Pi_1 - \Pi_2\|_{\mathrm{TV}} = 2 \sup_A |\Pi_1(A) - \Pi_2(A)|\,, \tag{4.2.1}$$

where the supremum runs over all measurable sets in Ω. In particular, the total variation distance between two probability measures is two if and only if they are mutually singular.

The factor two in this definition is convenient to include because this metric can be equivalently written as

$$\|\Pi_1 - \Pi_2\|_{\mathrm{TV}} = \sup \left\{ \left| \int_\Omega \varphi\, d\Pi_1 - \int_\Omega \varphi\, d\Pi_2 \right| \text{ s.t. } \begin{array}{c} \varphi : \Omega \to \mathbb{R}\,, \\ |\varphi(x)| \leq 1 \text{ for all } x \in \Omega \end{array} \right\}.$$

4.2.1. Harris Theorem. Harris Theorem states that if a Markov process admits a stochastic Lyapunov function such that its sublevel sets are 'small', then it is geometrically ergodic. More precisely, Harris Theorem applies to any Markov process $\mathsf{X}(t)$ on a Polish space Ω, with generator L, Markov semigroup P_t, and transition probabilities

$$\Pi_{t,x}(A) = \mathbb{P}(\mathsf{X}(t) \in A \mid \mathsf{X}(0) = x) \quad t > 0\,, \quad x \in \Omega\,,$$

that satisfies Assumptions 4.2 and 4.3 given below.

ASSUMPTION 4.2 (Infinitesimal Drift Condition). There exist a function $\Phi : \Omega \to \mathbb{R}^+$ and strictly positive constants w and k, such that

$$\mathsf{L}\Phi(x) \le \mathsf{k} - \mathsf{w}\,\Phi(x)\,, \tag{4.2.2}$$

for all $x \in \Omega$.

Assumption 4.2 implies that the associated semigroup P_t satisfies:

$$\mathsf{P}_t\Phi(x) \le e^{-\mathsf{w}t}\Phi(x) + \frac{\mathsf{k}}{\mathsf{w}}(1 - e^{-\mathsf{w}t})$$

for every $t \ge 0$ and for every $x \in \mathbb{R}^n$.

ASSUMPTION 4.3 (Associated 'Small Set' Condition). There exists a constant $\mathsf{a} \in (0,1)$ such that the sublevel set $\{z \in \Omega \mid \Phi(z) \le 2\mathsf{k}/\mathsf{w}\}$ is 'small' i.e.

$$\|\Pi_{t,x} - \Pi_{t,y}\|_{\mathrm{TV}} \le 2(1 - \mathsf{a}) \tag{4.2.3}$$

for all x, y satisfying $\Phi(x) \vee \Phi(y) \le 2\mathsf{k}/\mathsf{w}$, where the constants k and w are taken from Assumption 4.2.

REMARK 4.2.1. Assumption 4.3 is typically verified by proving that:

(1) There exists an *accessible point*: that is a $y \in \Omega$ such that

$$\Pi_{t,x}(\mathcal{O}_y) > 0 \tag{4.2.4}$$

for all $x \in \Omega$, for every neighborhood $\mathcal{O}_y$ of y, and for some $t \ge 0$.

(2) For every $t > 0$ the associated semigroup is *strongly Feller*:

$$\varphi \in B_b(\Omega) \implies \mathsf{P}_t\varphi \in C_b(\Omega). \tag{4.2.5}$$

If a Markov process has an accessible point and its semigroup is strongly Feller, then every compact set is small (including the sublevel set of the Lyapunov function mentioned in Assumption 4.3) and the process can have at most one invariant probability measure [**53, 123, 130**].

THEOREM 4.2.1 (Harris Theorem). *Consider a Markov process* $\mathsf{X}(t)$ *on* Ω *with generator* L *and transition probabilities* $\Pi_{t,x}$, *which satisfies Assumptions* 4.2 *and* 4.3. *Then* $\mathsf{X}(t)$ *possesses a unique invariant probability measure* Π, *and there exist positive constants* C *and* r *(both depending only on the constants* w, k *and* a *appearing in the assumptions) such that*

$$\|\Pi_{t,x} - \Pi\|_{\mathrm{TV}} \le \mathsf{C}\,\exp(-\mathsf{r}\,t)\,\Phi(x)\,,$$

for all $x \in \Omega$ *and for any* $t \ge 0$.

REMARK 4.2.2. It follows from (4.2.2) that:

$$\Pi(\Phi) \le \frac{\mathsf{k}}{\mathsf{w}}.$$

For further exposition and a proof of Harris Theorem in a general context, see the monograph [**123**] (or [**56**] for an alternative proof), and in the context of SDEs, see Section 2 and Appendix A of [**64**]. On a countable state space, the conditions in Theorem 4.2.1 become less onerous, and in fact, if the process is irreducible then all finite sets are small, in the sense of Assumption 4.3. In fact, it is sufficient to check irreducibility and an infinitesimal drift condition [**122**, see Theorem 7.1].

THEOREM 4.2.2 (Harris Theorem on a Countable State Space). *Consider an irreducible Markov process* X *on a countable state space* Ω *with generator* L *and transition probabilities* $\Pi_{t,x}$*, which satisfies the infinitesimal drift condition Assumption* 4.2. *Then* X *possesses a unique invariant probability measure* Π*, and there exist positive constants* C *and* r *(both depending only on the constants* w *and* k *appearing in the assumption) such that*

$$\|\Pi_{t,x} - \Pi\|_{\mathrm{TV}} \le \mathsf{C}\ \exp(-\mathsf{r}\ t)\ \Phi(x)\ ,$$

for all $x \in \Omega$ *and for any* $t \ge 0$.

4.2.2. Geometric Ergodicity of SDE Solution. The material here extends Theorem 2.12 in [**15**] to SDEs that are not necessarily reversible. We first show that a weak dissipativity condition on the drift implies that the generator L of the SDE (4.1.5) satisfies an infinitesimal drift condition.

LEMMA 4.2.3. *Suppose the drift field of the SDE* (4.1.5) *satisfies Assumption* 4.1. *Consider the function* $V : \mathbb{R}^n \to \mathbb{R}_+$ *defined as:*

$$V(x) = \exp\left(a|x|^{2m+2}\right) \quad \text{for any } 0 < a < \frac{\beta}{2(m+1)} \tag{4.2.6}$$

where m *and* β *are the constants appearing in Assumption* 4.1 *(A2) and Lemma* 4.1.2, *respectively. Then there exist positive constants* K *and* γ *such that*

$$LV(x) \le K - \gamma\, V(x) \tag{4.2.7}$$

holds for all $x \in \mathbb{R}^n$.

REMARK 4.2.3. Note that the Lyapunov function in (4.2.6) is defined in terms of the natural number m appearing in Assumption 4.1 (A2), which sets the leading order polynomial growth of the drift field μ. This Lyapunov function is sharp in the sense that it implies integrability of any observable that can be dominated by the exponential of the magnitude of the drift field. A discrete analog of this lemma is given in Lemma 4.2.6, and is a crucial ingredient to the subsequent analysis of the approximation.

PROOF. Set $\mathcal{E}(x) = x^T x$, $\Theta(u) = \exp(au^p)$, and $V = \Theta \circ \mathcal{E}$, where p is a real number and a is the positive real number appearing in Lemma 4.2.3. We assume that p is a free parameter at first. The proof then determines p such that the desired infinitesimal drift condition holds. From these definitions it follows that:

$$D\mathcal{E}(x)^T \eta = 2x^T \eta\ , \quad D^2\mathcal{E}(x)(\eta,\eta) = 2\eta^T \eta\ ,$$
$$\Theta'(u) = apu^{p-1}\Theta(u)\ , \quad \text{and} \quad \Theta''(u) = (a^2p^2u^{2p-2} + ap(p-1)u^{p-2})\Theta(u).$$

Hence,

$$\begin{aligned}\frac{L(\Theta\circ\mathcal{E})}{\Theta\circ\mathcal{E}} &= \frac{1}{\Theta\circ\mathcal{E}}\left((\Theta'\circ\mathcal{E})L\mathcal{E}+(\Theta''\circ\mathcal{E})|D\mathcal{E}|^2\right)\\ &= \frac{1}{\Theta\circ\mathcal{E}}\left((\Theta'\circ\mathcal{E})L\mathcal{E}+4(\Theta''\circ\mathcal{E})\mathcal{E}\right)\\ &= ap\mathcal{E}^{p-1}L\mathcal{E}+4a^2p^2\mathcal{E}^{2p-1}+4ap(p-1)\mathcal{E}^{p-1}\end{aligned}$$

Lemma 4.1.2 implies that

$$L\mathcal{E}\le -2\beta\mathcal{E}^{m+1}+2\alpha\mathcal{E}^{m+1/2}+2(\alpha+n).$$

Combining this inequality with the preceding equation yields,

$$\begin{aligned}\frac{L(\Theta\circ\mathcal{E})}{\Theta\circ\mathcal{E}} &\le 2ap(2ap\mathcal{E}^{2p-1}-\beta\mathcal{E}^{p+m})\\ &\quad + (4ap(p-1)+2ap(\alpha+n)+2ap\alpha\mathcal{E}^{m+1/2})\mathcal{E}^{p-1}.\end{aligned}$$

Given the hypothesis on a, the negative term $-\beta\ \mathcal{E}^{p+m}$ in this upper bound dominates over the leading order positive term $2\ a\ p\ \mathcal{E}^{2p-1}$ provided that $p\le m+1$. Hence, the largest p can be is $m+1$. Substituting this value of p yields:

$$\frac{L(\Theta\circ\mathcal{E})}{\Theta\circ\mathcal{E}}\le 2a(m+1)\underbrace{(2a(m+1)-\beta)}_{<\,0\text{ by hypothesis on }a}\ \mathcal{E}^{2m+1}+\text{lower order in }\mathcal{E}\text{ terms.}$$

Hence, for any $\gamma>0$, there exists $E^\star$ such that:

$$\frac{L(\Theta\circ\mathcal{E})}{\Theta\circ\mathcal{E}}\le-\gamma\quad\text{for all }|x|^2>E^\star.$$

Accordingly, a suitable constant K in (4.2.7) is:

$$K=\sup_{|x|^2\le E^\star}|L(\Theta\circ\mathcal{E})+\gamma(\Theta\circ\mathcal{E})|$$

which is bounded since the set $\{|x|^2\le E^\star\}$ is compact and $\Theta\circ\mathcal{E}$ is smooth. □

LEMMA 4.2.4. *Suppose the drift field of the SDE* (4.1.5) *satisfies Assumption* 4.1. *For every* $t>0$ *and* $E>0$, *there exists* $\epsilon\in(0,1)$ *such that*

$$\|\Pi_{t,x}-\Pi_{t,y}\|_{\mathrm{TV}}\le 2(1-\epsilon)\ , \tag{4.2.8}$$

for all x,y *satisfying* $V(x)\vee V(y)\le E$.

PROOF. It follows from the assumed ellipticity of the equations and the assumed regularity of the drift field, that there exists a function $p(t,x,y)$ continuous in all of its arguments (for $t>0$) such that the transition probabilities are given by $\Pi_{t,x}(dy)=p(t,x,y)\,dy$. Furthermore, p is strictly positive (see, e.g., Prop. 2.4.1 of [**23**]). Hence, by the compactness of the set $\{x\ :\ V(x)\le E\}$, one can find a probability measure η on $\mathbb{R}^n$ and a constant $\epsilon>0$ such that,

$$\Pi_{t,x}>\epsilon\eta$$

for any x satisfying $V(x)\le E$. This condition implies the following transition probability $\tilde{P}_t$ on $\mathbb{R}^n$ is well-defined:

$$\tilde{\Pi}_{t,x}=\frac{1}{1-\epsilon}\Pi_{t,x}-\frac{\epsilon}{1-\epsilon}\eta$$

for any x satisfying $V(x)\le E$. Therefore,

$$\|\Pi_{t,x}-\Pi_{t,y}\|_{\mathrm{TV}}=(1-\epsilon)\|\tilde{\Pi}_{t,x}-\tilde{\Pi}_{t,y}\|_{\mathrm{TV}}$$

for all x, y satisfying $V(x) \vee V(y) \leq E$. Since the TV norm is always bounded by 2, one obtains the desired result. □

With Lemmas 4.2.3 and 4.2.4, we can now invoke Harris Theorem 4.2.1 to obtain the following geometric ergodicity result for the SDE solution.

THEOREM 4.2.5. *Suppose the drift field of the SDE* (4.1.5) *satisfies Assumption* 4.1. *Let* $V(x)$ *be the Lyapunov function from Lemma* 4.2.3. *Then, the SDE admits a unique invariant probability measure* Π. *Moreover, there exist positive constants* λ *and* C *such that*

$$\|\Pi_{t,x} - \Pi\|_{\mathrm{TV}} \leq C \exp(-\lambda t) V(x) \tag{4.2.9}$$

for all $t \geq 0$ *and for all* $x \in \mathbb{R}^n$.

PROOF. According to Lemma 4.2.3, for any $a \in (0, \beta(2(m+1))^{-1})$, $V(x)$ is a Lyapunov function for the SDE. Moreover, by Lemma 4.2.4 it satisfies a small set condition on every sublevel set of V. Hence, Harris Theorem implies that the SDE solution has a unique invariant probability measure Π and that (4.2.9) holds. □

4.2.3. Geometric Ergodicity of Approximation. Here we show that the Markov process induced by the generator Q in (4.1.8) is geometrically ergodic. The main ingredient in this proof is a discrete analog of Lemma 4.2.3, which shows that the approximation has a stochastic Lyapunov function.

LEMMA 4.2.6. *Suppose the drift field of the SDE satisfies Assumption* 4.1. *Let*

$$V(x) = \exp\left(a|x|_{2m+2}^{2m+2}\right) \quad \text{for any } 0 < a < \frac{\beta_0}{2(m+1)} \tag{4.2.10}$$

where m *and* β_0 *are the constants appearing in Assumption* 4.1 *(A2). Then, there exist positive constants* h_c, K *and* γ *(uniform in* h*) such that*

$$QV(x) \leq K - \gamma V(x) \tag{4.2.11}$$

holds for all $x \in \mathbb{R}^n$ *and* $h < h_c$.

REMARK 4.2.4. Note that the stochastic Lyapunov function in (4.2.10) differs from the one defined in equation (4.2.6) of Theorem 4.2.3. In particular, the Lyapunov function in (4.2.6) uses the 2-norm whereas the one in (4.2.10) uses the $(2m+2)$-norm. As we will see below, using the $(2m+2)$-norm in (4.2.10) is convenient for the analysis and makes the proof more transparent, particularly since the jump directions in Q are along the coordinate axes. However, since all norms in $\mathbb{R}^n$ are equivalent, this difference is minor, and in particular, does not restrict the type of functions we can bound by (4.2.10).

The following lemma is repeatedly used in the proof of Lemma 4.2.6.

LEMMA 4.2.7. *Suppose that* $\mathsf{a}, \mathsf{b} \geq 0$ *and* $\mathsf{k} \in \mathbb{N}$. *Then we have the bound:*

$$\mathsf{a}^{\mathsf{k}} - \mathsf{b}^{\mathsf{k}} \leq \mathsf{k}\,(\mathsf{a} - \mathsf{b})\,\mathsf{a}^{\mathsf{k}-1}. \tag{4.2.12}$$

PROOF. Note that the bound holds if $\mathsf{a} = 0$ for any $\mathsf{b} \geq 0$ and for any $\mathsf{k} \in \mathbb{N}$. For $\mathsf{a} > 0$, the bound comes from Bernoulli's inequality

$$\left(\frac{\mathsf{b}}{\mathsf{a}}\right)^{\mathsf{k}} \geq 1 + \mathsf{k}\,(\frac{\mathsf{b}}{\mathsf{a}} - 1). \tag{4.2.13}$$

Indeed, recall that Bernoulli's inequality holds for all $\mathsf{b}/\mathsf{a} \geq 0$ and $\mathsf{k} \in \mathbb{N}$, with equality if and only if $\mathsf{a} = \mathsf{b}$ or $\mathsf{k} = 1$. Since $\mathsf{a} > 0$, (4.2.13) implies (4.2.12). □

Here is the proof of Lemma 4.2.6. We stress that this proof heavily relies on the fact that the spatial step size of the approximation is uniformly bounded.

PROOF. Apply the generator Q in (4.1.8) to $V(x)$ to obtain:

$$\frac{QV(x)}{V(x)} = \sum_{i=1}^{n} (A_i^+(x) + A_i^-(x)) \tag{4.2.14}$$

where we have defined:

$$A_i^{\pm}(x) = \frac{1}{h^2} \exp\left(\pm \frac{h}{2}\mu_i(x)\right) \left(\frac{V_i^{\pm}(x)}{V(x)} - 1\right) \quad \text{and} \quad V_i^{\pm}(x) = V(x \pm e_i h)$$

for $1 \le i \le n$.

Step 1: Bound $A_i^{\pm}$. Lemma 4.2.7 with

$$\mathsf{a} = (x_i \pm h)^2 \,, \quad \mathsf{b} = x_i^2 \,, \quad \text{and} \quad \mathsf{k} = m+1$$

implies that:

$$\begin{aligned} \frac{V_i^{\pm}(x)}{V(x)} &= \exp\left(a((x_i \pm h)^2)^{m+1} - a(x_i^2)^{m+1} \right) \\ &\le \exp\left(a(m+1)(\pm 2 x_i h + h^2)(x_i \pm h)^{2m} \right) \\ &\le \exp\left(\pm 2a(m+1) x_i^{2m+1} h + R_i^{\pm} \right) \end{aligned}$$

where for the sake of readability we have introduced:

$$R_i^{\pm}(x) = a(m+1) x_i^{2m} h^2 + a(m+1)(\pm 2 x_i h + h^2) \left(((x_i \pm h)^2)^m - (x_i^2)^m \right) .$$

Two more applications of Lemma 4.2.7 show that

$$|((x_i \pm h)^2)^m - (x_i^2)^m| \le m |(x_i \pm h)^2 - x_i^2| (x_i^2 \vee (x_i \pm h)^2)^{m-1}$$

from which it follows that

$$\begin{aligned} R_i^{\pm}(x) &\le a(m+1) x_i^{2m} h^2 + am(m+1)(\pm 2 x_i h + h^2)^2 (2x_i^2 + 2h^2)^{m-1} \\ &\le a(m+1) x_i^{2m} h^2 + am(m+1)(8 x_i^2 h^2 + 2h^4)(2x_i^2 + 2h^2)^{m-1} \\ &\le (C_1 x_i^{2m} + C_0) h^2 \end{aligned}$$

for some $C_0, C_1 \ge 0$, for all $x \in \mathbb{R}^n$ and for any $h \in (0,1)$. Thus, we obtain the following bound:

$$\begin{aligned} A_i^{\pm}(x) \le \frac{1}{h^2} &\exp\left(\pm \frac{h}{2} \mu_i \right) \\ &\times \left(\exp\left(\pm 2a(m+1) x_i^{2m+1} h + (C_1 x_i^{2m} + C_0) h^2 \right) - 1 \right) \end{aligned} \tag{4.2.15}$$

Step 2: Merge bounds. Combine (4.2.15) with (4.2.14) to obtain:

$$\frac{QV(x)}{V(x)} \le \phi(h,x) = \sum_{i=1}^{n} \phi_i(h,x) \tag{4.2.16}$$

where we have introduced:

$$\phi_i(s,x) = \frac{2}{s^2} \Big(\exp(c_i(x) s^2) \cosh(a_i(x) s) - \cosh(b_i(x) s) \Big) \tag{4.2.17}$$

and the following functions of position only:

$$\begin{cases} c_i(x) = C_1 x_i^{2m} + C_0 \\ b_i(x) = \frac{1}{2}\mu_i(x) \\ a_i(x) = 2a(m+1)x_i^{2m+1} + b_i(x) \end{cases} \tag{4.2.18}$$

Note that $c_i(x)$ only depends on the ith component of x, and for $j \neq i$, $b_i(x)$ and $a_i(x)$ depend on the jth-component of x through $\mu_i(x)$.

Step 3: Determine Leading Order Behavior of $\phi(h,x)$. The order of magnitude of the terms in $\phi(s,x)$ as $|x| \to \infty$ is assessed here. Note that the weak dissipativity and minimal growth conditions in Assumptions 4.1 (A3) and (A4) respectively, imply that

$$\begin{aligned} \phi_i(s,x) &\le \frac{2}{s^2}\Bigg(\exp(c_i(x)s^2)\cosh\left(2a(m+1)|x_i|^{2m+1} + \frac{\mu_i(x)}{2}\operatorname{sign}(x_i)\right) \\ &\qquad - \cosh\left(\frac{\beta_0}{2}|x_i|^{2m+1} - \frac{\alpha_0}{2}(1+|x|^{2m})\right)\Bigg) \\ &\le \frac{2}{s^2}\Bigg(\exp(c_i(x)s^2)\cosh\left(\left(2a(m+1) - \frac{\beta_0}{2}\right)|x_i|^{2m+1} + \frac{\alpha_0}{2}(1+|x|^{2m})\right) \\ &\qquad - \cosh\left(\frac{\beta_0}{2}|x_i|^{2m+1} - \frac{\alpha_0}{2}(1+|x|^{2m})\right)\Bigg) \end{aligned} \tag{4.2.19}$$

where in the first step we used the identity $\cosh(\operatorname{sign}(x_i)f(x)) = \cosh(f(x))$ for any $f : \mathbb{R}^n \to \mathbb{R}$. Our hypothesis on a implies that $2a(m+1) - \beta_0/2 < \beta_0/2$, and hence, the second term in (4.2.19) dominates over the first term as $|x_i| \to \infty$. Summing these inequalities over i yields a bound on $\phi(s,x)$. As $|x| \to \infty$ (or $\max(|x_1|, \cdots, |x_n|) \to \infty$), the negative terms in this bound on $\phi(s,x)$ dominate over the positive terms. Thus, the weak dissipativity and minimal growth condition on the drift field are the source of dissipation in Q. In contrast to the proof of Lemma 4.2.3, here we require a minimal growth condition in order to show that Q has a dissipative effect. However, as shown in Example 1, this additional condition is not too restrictive and is satisfied by the drift fields we have in mind.

Step 4: Use monotonicity. To obtain a bound that is uniform in the spatial step size h, we use a monotonicity argument. We propose to upper bound the right-hand-side of (4.2.16) by the limit of $\phi(h,x)$ as $h \to 0$. Indeed, by l'Hôpital's Rule this limit is given by:

$$\begin{aligned} \lim_{s\to 0}\phi(s,x) &= \sum_{i=1}^n a_i^2 - b_i^2 + 2c_i \\ &= \sum_{i=1}^n 2a(m+1)x_i^{2m}(2a(m+1)x_i^{2m+2} + \mu_i x_i) + 2c_i \\ &= \sum_{i=1}^n 2a(m+1)(2a(m+1) - \beta_0)x_i^{4m+2} + \cdots \\ &\le 2a(m+1)\underbrace{(2a(m+1) - \beta_0)}_{<0 \text{ by hypothesis on } a}|x|_{4m+2}^{4m+2} + \text{lower order in } |x| \text{ terms} \end{aligned} \tag{4.2.20}$$

where in the third step we used Assumption 4.1 (A3) to bound $\mu_i x_i = \mu_i \operatorname{sign}(x_i)|x_i|$. Thus, this upper bound can be made arbitrarily large and negative if $|x|$ is large enough. Another application of l'Hôpital's Rule shows that

$$\lim_{s \to 0} \frac{\partial \phi}{\partial s}(s,x) = 0 \quad \forall x \in \mathbb{R}^n \tag{4.2.21}$$

and hence, the point $s = 0$ is a critical point of $\phi(s,x)$. To determine the properties of this critical point, we apply l'Hôpital's Rule and the minimal growth condition in Assumption 4.1 (A4) to obtain:

$$\begin{aligned}
\lim_{s \to 0} \frac{\partial^2 \phi}{\partial s^2}(s,x) &= \sum_{i=1}^n \frac{1}{6}(a_i^4 - b_i^4 + 12 a_i^2 c_i + 12 c_i^2)\ , \\
&= \sum_{i=1}^n \frac{1}{6}(a_i^2 - b_i^2)(a_i^2 + b_i^2) + a_i^2 c_i + c_i^2 \\
&\le \sum_{i=1}^n \frac{1}{3} a(m+1)(2a(m+1) - \beta_0) x_i^{4m+2}(a_i^2 + b_i^2) + \cdots \\
&\le \sum_{i=1}^n \frac{1}{12} a(m+1)(2a(m+1) - \beta_0) x_i^{4m+2} \mu_i^2 + \cdots \\
&\le \sum_{i=1}^n \frac{\beta_0^2}{24} a(m+1)(2a(m+1) - \beta_0) x_i^{8m+4} + \cdots \\
&\le \frac{\beta_0^2}{24} a(m+1) \underbrace{(2a(m+1) - \beta_0)}_{<0 \text{ by hypothesis on } a} |x|_{8m+4}^{8m+4} + \text{lower order in } |x| \text{ terms}
\end{aligned} \tag{4.2.22}$$

which is negative if $|x|$ is large enough. By the pointwise second derivative rule, $\lim_{s \to 0} \phi(s,x)$ is a local maximum of the function $\phi(s,x)$, and therefore, the function $\phi(s,x)$ is strictly decreasing in a neighborhood of $s = 0$. Thus, for all x sufficiently large and for s sufficiently small, we may use $\lim_{s \to 0} \phi(s,x)$ in (4.2.20) to bound $\phi(s,x)$.

Step 5: Conclude Proof. We have shown that there exist positive constants h_c, $R^\star$ and γ (all independent of h) such that:

$$\frac{QV(x)}{V(x)} < -\gamma \quad \text{for all } x \in \mathbb{R}^n \text{ satisfying } |x| > R^\star \tag{4.2.23}$$

and for all $h < h_c$. To finish, define K in (4.2.11) to be:

$$K = \sup_{|x| \le R^\star} |LV(x) + \gamma V(x)| + \sup_{0<h<h_c} \sup_{|x| \le R^\star} |(Q-L)V(x)|.$$

The first term of K is independent of h and bounded, since $LV(x) + \gamma V(x)$ is smooth and the set $\{|x| \le R^\star\}$ is compact. According to Lemma 4.4.1 below, the difference $|(Q-L)V(x)|$ is $\mathcal{O}(h^2)$, involves four derivatives of $V(x)$ and the drift field μ. Since this function is continuous and the set $\{|x| \le R^\star\}$ is compact, this term can be bounded uniformly in h. ☐

We are now in position to invoke the countable state space version of Harris Theorem 4.2.2, which implies geometric ergodicity of the approximation.

THEOREM 4.2.8 (Geometric Ergodicity of Approximation). *Suppose the drift field of the underlying SDE* (4.1.5) *satisfies Assumption* 4.1. *Then the Markov process induced by Q has a unique invariant probability measure Π^h supported on the grid S_h. Furthermore, there exist positive constants λ and C such that*

$$\|\Pi_{t,x}^h - \Pi^h\|_{\mathrm{TV}} \le C\, V(x)\, e^{-\lambda t} \tag{4.2.24}$$

for all $t \ge 0$ and all $x \in S_h$.

Recall that on a countable state space the TV distance between probability measures is equivalent to the ℓ_1 distance between their probability densities. Thus, (4.2.24) is a bound on the ℓ_1 distance between the transition and stationary probability densities of the approximation.

PROOF. The Markov process induced by Q is irreducible. Indeed, at every point on the grid, the transition rate to nearest neighbors is strictly positive. Thus, for any two grid points x and y, it is possible to find a finite sequence of jumps from x to y, and for any $t > 0$, there is a strictly positive probability that this finite sequence of jumps will occur. Moreover, the generator Q satisfies an infinitesimal drift condition according to Lemma 4.2.6. Hence, Theorem 4.2.2 implies that (i) the process possesses a stationary distribution, and (ii) the transition probability of the process converges to this stationary distribution geometrically fast in the TV metric. □

REMARK 4.2.5. Let $\|\cdot\|_V$ denote the V-weighted supremum norm defined as:

$$\|f\|_V = \sup_{x \in S_h} \frac{|f(x)|}{1 + V(x)}$$

where $f : S_h \to \mathbb{R}$ and $V : \mathbb{R}^n \to \mathbb{R}$ is a Lyapunov function of Q from Lemma 4.2.6. Define the following set of grid functions:

$$\ell_V = \{g : S_h \to \mathbb{R} \mid \|g\|_V < \infty\}.$$

The conditions of Theorem 4.2.8 imply exponential convergence in this function space:

$$\|P_t^h f - \Pi^h(f)\|_V \le C\|f - \Pi^h(f)\|_V e^{-\tilde{\lambda} t}\,, \tag{4.2.25}$$

for some $\tilde{\lambda} > 0$ and $C > 0$, for all $t \ge 0$ and for all $f \in \ell_V$ [**85, 86, 135**]. Here we used the notation:

$$\Pi^h(f) = \sum_{x \in S_h} f(x) \nu^h(x).$$

REMARK 4.2.6. Consider the Hilbert space

$$\ell^2(\Pi^h) = \{f : S_h \to \mathbb{R} \mid \sum_{x \in S_h} |f(x)|^2 \nu^h(x) < \infty\}$$

with inner product:

$$\langle f, g \rangle_{\ell^2(\Pi^h)} = \sum_{x \in S} f(x) g(x) \nu^h(x).$$

If Q is self-adjoint, in the sense that

$$\langle f, Qg \rangle_{\ell^2(\Pi^h)} = \langle Qf, g \rangle_{\ell^2(\Pi^h)} \quad \text{for all } f, g \in \ell_2(\Pi^h)$$

then (4.2.24) holds if and only if there exists a $\tilde{\lambda} > 0$ such that:

$$\|P_t^h f - \Pi^h(f)\|_{\ell^2(\Pi^h)} \le C\|f - \Pi^h(f)\|_{\ell^2(\Pi^h)} e^{-\tilde{\lambda} t}\,, \tag{4.2.26}$$

for all $t \geq 0$ and for all $f \in \ell_2(\Pi^h)$ [**24, 135**]. We stress that this equivalence does not hold in general if the Markov process is not self-adjoint.

REMARK 4.2.7. Let $\sigma(Q)$ denote the spectrum of Q: the set of all points $\lambda \in \mathbb{C}$ for which $Q - \lambda I$ is not invertible. If Q possesses a spectral gap in $\ell^2(\Pi^h)$, i.e., a gap between its zero eigenvalue and the eigenvalue of second smallest magnitude, then its transition probabilities converge exponentially fast in $\ell^2(\Pi^h)$. Indeed, by the spectral mapping theorem,

$$\sigma(P_t^h) = \{e^{t\lambda} \mid \lambda \in \sigma(Q)\}$$

and hence, a spectral gap in Q implies that P_t^h has a gap between its unit eigenvalue and its second largest eigenvalue [**24, 135**].

4.3. Properties of Realizations

4.3.1. Basic Properties of SDE Solution. Here we review some basic properties of the SDE solution under Assumption 4.1 on the drift field.

LEMMA 4.3.1. *Suppose the drift field of the SDE* (4.1.5) *satisfies Assumption* 4.1. *For every* $\varphi : \mathbb{R}^n \to \mathbb{R}$ *satisfying* $|\varphi| \leq V + C_0$ *for some* $C_0 > 0$, *we have that:*

$$|\mathbb{E}_x \varphi(Y(t))| \leq V(x) + \frac{K}{\gamma} + C_0$$

for all $t \geq 0$ *and for all* $x \in \mathbb{R}^n$. *Here* K *and* γ *are taken from Lemma* 4.2.3.

As a consequence of this lemma, all moments of the SDE solution are bounded uniformly in time.

PROOF. Since φ is dominated by V, this result follows directly from Lemma 4.2.3. Indeed,

$$|P_t\varphi(x)| \leq P_t|\varphi(x)| \leq P_t V(x) + C_0 \leq e^{-\gamma t} V(x) + \frac{K}{\gamma} + C_0.$$

□

The next lemma bounds the mean-squared displacement of the SDE solution.

LEMMA 4.3.2. *Suppose the drift field of the SDE* (4.1.5) *satisfies Assumption* 4.1. *For all* $t \geq 0$ *and for all* $x \in \mathbb{R}^n$, *the displacement of the SDE solution* Y *satisfies:*

$$\mathbb{E}_x |Y(t) - x|^2 \leq 2\, t^2 \left(V(x) + \frac{K}{\gamma} + C_0 \right) + 4\, n\, t$$

where K, γ *and* C_0 *are taken from Lemma* 4.3.1.

PROOF. Jensen's inequality applied to the convex function $x \mapsto x^2$ and the Cauchy-Schwarz inequality imply that

$$\begin{aligned} |Y(t) - x|^2 &\leq 2 \left| \int_0^t \mu(Y(s)) ds \right|^2 + 4 \left| \int_0^t dW(s) \right|^2 \\ &\leq 2t \int_0^t |\mu(Y(s))|^2 ds + 4 \left| \int_0^t dW(s) \right|^2. \end{aligned}$$

Take expectations, use the maximal polynomial growth of μ from Assumption 4.1 (A2), apply Lemma 4.3.1, and invoke the Itō isometry to derive the desired result.

□

The one-sided Lipschitz property of the drift field from Assumption 4.1 (A1) (see also Lemma 4.1.1) implies that the SDE solution is almost surely Lipschitz continuous with respect to its initial condition as the next lemma states.

LEMMA 4.3.3. *Suppose the drift field of the SDE (4.1.5) satisfies Assumption 4.1. For $s \le t$, let $Y_{t,s}(x)$ denote the evolution map of the SDE, which satisfies $Y_{s,s}(x) = x$ and $Y_{t,s} \circ Y_{s,r}(x) = Y_{t,r}(x)$ for $r \le s \le t$. For all $t \ge s$ and for all $x, y \in \mathbb{R}^n$, this evolution map satisfies:*

$$|Y_{s+t,s}(x) - Y_{s+t,s}(y)|^2 \le |x-y|^2 \exp(2\, C_L^\mu\, t) \quad a.s.$$

where C_L^μ is the one-sided Lipschitz constant of the drift field from Assumption 4.1 (A1).

PROOF. By Itō-Taylor's formula:

$$\begin{aligned}
&|Y_{s+t,s}(x) - Y_{s+t,s}(y)|^2 \\
&\quad \le |x-y|^2 + 2\int_0^t \langle Y_{s+r,s}(x) - Y_{s+r,s}(y), \mu(Y_{s+r,s}(x)) - \mu(Y_{s+r,s}(y))\rangle\, dr \\
&\quad \le |x-y|^2 + 2\, C_L^\mu \int_0^t |Y_{s+r,s}(x) - Y_{s+r,s}(y)|^2\, dr
\end{aligned}$$

where we used the one-sided Lipschitz property of the drift field with Lipschitz constant C_L^μ. An application of Gronwall's Lemma implies the desired result. □

We now turn to estimates related to the semigroup of the SDE solution.

LEMMA 4.3.4. *Suppose Assumption 4.1 holds. For all $t > 0$, for all $x, y \in \mathbb{R}^n$, and for all $\varphi \in B_b(\mathbb{R}^n)$, there exists $C > 0$ such that*

$$|P_t\varphi(x) - P_t\varphi(y)| \le C\,(1 \wedge t)^{-1/2}\, \|\varphi\|_\infty\, |y-x|.$$

We stress that the test function in Lemma 4.3.4 is not necessarily continuous. Assuming that $t > 0$, this lemma shows that P_t regularizes test functions, and specifically that the function $P_t\varphi$ is Lipschitz continuous for any $\varphi \in B_b(\mathbb{R}^n)$. Since $P_t\varphi$ is also bounded, it follows that the semigroup P_t is strongly Feller, see Remark 4.2.1 for a definition of the strong Feller property. Let us stress that this property of P_t relies on the generator L of the SDE being of elliptic type. For a detailed proof of this result that only requires local Lipschitz continuity and a polynomial growth condition on the drift field, we refer to §2.3.1 of [**23**].

Now we make use of the fact that the drift field is four times differentiable and these derivatives satisfy a maximal polynomial growth condition.

LEMMA 4.3.5. *Suppose Assumption 4.1 holds. For any $p \ge 1$, for any $1 \le j \le 4$, for any $t \ge s$, and for all $x \in \mathbb{R}^n$, the evolution map $Y_{t,s}(x)$ is four times mean-square differentiable and*

$$\sup_{x\in\mathbb{R}^n} \mathbb{E}\|D^j Y_{t,s}(x)\|^p \le C_p(t)$$

where $C_p(t) > 0$.

See Theorem 1.3.6 of [**23**] for a detailed proof of Lemma 4.3.5 under the stated assumptions. We apply this result in the next lemma.

LEMMA 4.3.6. *Suppose Assumption* 4.1 *holds. For any* $1 \leq j \leq 4$, *for any* $t \geq 0$, *and for any* $\varphi \in C_b^4(\mathbb{R}^n)$, *the function* $P_t\varphi$ *is* j *times differentiable and there exists* $C(t) > 0$ *such that*

$$\|P_t\varphi\|_j \leq C(t)\, \|\varphi\|_j$$

where $\|\cdot\|_j$ *is the norm introduced in* (4.1.3).

In short, this lemma states that P_t maps $C_b^4(\mathbb{R}^n)$ into itself for any $t \geq 0$.

PROOF. For any $1 \leq j \leq 4$, differentiate j times with respect to x the probabilistic representation of the semigroup:

$$P_t\varphi(x) = \mathbb{E}\varphi(Y_{t,0}(x))$$

and use Lemma 4.3.5 to bound these derivatives of $Y_{t,0}(x)$. For example, for $j = 1$ the chain rule implies that:

$$DP_t\varphi(x)(\eta) = \mathbb{E}D\varphi(Y_{t,0}(x))(DY_{t,0}(x)(\eta))$$

for any $\eta \in \mathbb{R}^n$, for all $x \in \mathbb{R}^n$ and for all $t \geq 0$. Hence,

$$\begin{aligned} |DP_t\varphi(x)(\eta)| &\leq |\mathbb{E}D\varphi(Y_{t,0}(x))(DY_{t,0}(x)(\eta))| \\ &\leq \mathbb{E}|D\varphi(Y_{t,0}(x))(DY_{t,0}(x)(\eta))| \\ &\leq \mathbb{E}\,|D\varphi(Y_{t,0}(x))|\,\|DY_{t,0}(x)\|\,|\eta| \\ &\leq C(t)|\eta| \left(\sup_{x\in\mathbb{R}^n} |D\varphi(x)| \right) \end{aligned}$$

for some $C(t) > 0$. Here we used the fact that φ has a bounded derivative, Lemma 4.3.5, and the dominated convergence theorem to interchange differentiation and expectation. Since this upper bound is uniform in x, we obtain the desired estimate for $j = 1$. □

REMARK 4.3.1. As an alternative to this probabilistic proof of Lemma 4.3.6 which is due to [**23**], an analytic approach to bounding the derivatives of $P_t\varphi$ is available in [**9, 107**]. This approach is based on approximating $P_t\varphi$ by the solutions to Cauchy problems on compact spaces and assumes conditions which are similar to Assumption 4.1.

4.3.2. A Martingale Problem. The process X induced by Q has the following classification.

- X is right-continuous in time with left limits (càdlàg).
- X is adapted to the filtration associated to the driving noise.
- X is a pure jump process, since the process only changes its state by jumps.
- X has the Markov property, since conditional on $X(t_0) = x$, the time to the next jump is exponentially distributed with parameter that depends only on the state x; and, the jump itself has a probability distribution that only depends on the state x too.

As such the process admits the following semimartingale representation:

$$X(t) = X(0) + \int_0^t \mu^h(X(s))ds + \overbrace{\mathcal{M}(t)}^{\text{local martingale}}. \tag{4.3.1}$$

The stochastic equation introduced earlier in (4.1.12) is shorthand for this integral form of X. Let us prove that the local martingale in this representation is a global martingale. The next Lemma is useful for this purpose.

LEMMA 4.3.7. *Suppose Assumption* 4.1 *holds on the drift field, let* $V(x)$ *be the stochastic Lyapunov function from Lemma* 4.2.6, *and let* $\varphi : \mathbb{R}^n \to \mathbb{R}$. *Consider the local martingale:*

$$M^{\varphi}(t) = \varphi(X(t)) - \varphi(x) - \int_0^t Q\varphi(X(s))ds, \quad t \geq 0\,, \quad X(0) = x.$$

If φ *satisfies the growth condition:*

$$Q\varphi^2(x) - 2\varphi(x)Q\varphi(x) \leq V(x) + C \quad \textit{for all } x \in \mathbb{R}^n \tag{4.3.2}$$

for some $C > 0$, *then there exists a positive constant* h_c *such that* $M^{\varphi}(t)$ *is a global martingale for all* $h < h_c$.

It is not easy to relax condition (4.3.2), since it is based on a sharp stochastic Lyapunov function for the process.

PROOF. The idea of the proof is to show that the local martingale $M^{\varphi}(t)$ is dominated by an integrable random variable, namely the maximum process $M^*(t) = \sup_{0 \leq s \leq t} |M^{\varphi}(s)|$. (Recall that a local martingale dominated by an integrable random variable is a uniformly integrable global martingale [**83**, **131**, **133**].) To show that the maximum process is integrable, we apply the Burkholder-Davis-Gundy inequality with parameter $p = 1$, which implies that there exists a positive constant C_1 such that

$$\mathbb{E}_x M^*(t) \leq C_1 \mathbb{E}_x \sqrt{\langle M^{\varphi}, M^{\varphi}\rangle(t)} \leq C_1 \sqrt{\mathbb{E}_x \langle M^{\varphi}, M^{\varphi}\rangle(t)}$$

where we used Jensen's inequality with respect to the (concave) square root function.

Since the compensator of the quadratic variation of $M^{\varphi}(t)$ is given by (4.1.11), the hypothesis (4.3.2) implies that there exist positive constants C_2 and C_3 such that

$$\mathbb{E}_x M^*(t) \leq C_1 \sqrt{\mathbb{E}_x \int_0^t (V(X(s)) + C_2) ds} \leq (V(x) + C_3)t$$

Here we used Lemma 4.2.6, which implies that $V(x)$ is a stochastic Lyapunov function for X. Hence, $M^*(t)$ is integrable and $M^{\varphi}(t)$ is a uniformly integrable global martingale, for all $t \geq 0$. □

Thus Dynkin's formula (4.1.10) holds for any function satisfying the growth condition (4.3.2). The next lemma illustrates that this class of functions includes locally Lipschitz functions. Since the identity map is globally Lipschitz, this lemma enables studying solutions to the stochastic equation (4.3.1) at any $t \geq 0$.

LEMMA 4.3.8. *Suppose Assumption* 4.1 *holds on the drift field. Suppose* $\varphi : \mathbb{R}^n \to \mathbb{R}$ *is a locally Lipschitz continuous function*

$$|\varphi(y) - \varphi(x)| \leq L_{\varphi}(x, y)|y - x| \quad \forall\, x, y \in \mathbb{R}^n$$

with local Lipschitz constant $L_{\varphi}(x, y)$ *that satisfies the following condition*

$$L_{\varphi}(x, y) \leq C_{\varphi}(1 + \exp(|x|^{2m+1}) + \exp(|y|^{2m+1})) \quad \forall\, x, y \in \mathbb{R}^n$$

for some $C_{\varphi} > 0$. *Then* φ *satisfies* (4.3.2).

REMARK 4.3.2. In §4.5, we show that for any $\varphi \in C_b^4(\mathbb{R}^n)$, $P_t^h\varphi$ is accurate in approximating $P_t\varphi$ on both finite and infinite time intervals. Since functions in this space are Lipschitz continuous, Lemma 4.3.8 implies that Dynkin's formula holds for any $\varphi \in C_b^4(\mathbb{R}^n)$.

PROOF. Let $G^\varphi(x) = Q\varphi^2(x) - 2\varphi(x)Q\varphi(x)$. Use (4.1.8) to write

$$\begin{aligned} G^\varphi(x) = \sum_{i=1}^n \frac{1}{h^2} \exp\left(\mu_i \frac{h}{2}\right) (\varphi(x+he_i) - \varphi(x))^2 \\ + \frac{1}{h^2} \exp\left(-\mu_i \frac{h}{2}\right) (\varphi(x-he_i) - \varphi(x))^2 \end{aligned} \tag{4.3.3}$$

Since φ is locally Lipschitz continuous by hypothesis, we have that

$$\begin{aligned} G^\varphi(x) &\le \sum_{i=1}^n C_\varphi^2 \left(1 + \exp(|x|^{2m+1}) + \exp(|x+e_i h|^{2m+1})\right)^2 \exp\left(\mu_i \frac{h}{2}\right) \\ &\qquad + C_\varphi^2 \left(1 + \exp(|x|^{2m+1}) + \exp(|x-e_i h|^{2m+1})\right)^2 \exp\left(-\mu_i \frac{h}{2}\right) \\ &\le 2\, C_1 \sum_{i=1}^n \left(1 + \exp(|x|^{2m+1})\right) \cosh\left(\mu_i \frac{h}{2}\right) \\ &\le 2\, n\, C_1 \left(1 + \exp(|x|^{2m+1})\right) \cosh\left(C_P^\mu (1 + |x|^{2m+1}) \frac{h}{2}\right) \\ &\le V(x) + C_2 \end{aligned}$$

where C_1 and C_2 are positive constants. Thus, $\varphi(x)$ satisfies the growth condition (4.3.2), as required. Note that in the input to the cosh function, we used the bound:

$$|\mu_i(x)| \le |\mu(x)| \le C_P^\mu (1 + |x|^{2m+1}) \quad \forall\, x \in \mathbb{R}^n$$

which follows from the maximal polynomial growth condition in Assumption 4.1 (A2). Also in the last step, to obtain the bound by the stochastic Lyapunov function introduced in (4.2.10) we used the inequality $|x|^{2m+1} \le |x|^{2m+2}$ which is valid for $|x| > 1$, and the fact that all norms on $\mathbb{R}^n$ are equivalent, see Remark 4.2.4. □

As a tool to analyze realizations, we quote l'Hôpital's monotonicity principle as a lemma (without proof).

LEMMA 4.3.9. *Let $f, g : [a, b] \to \mathbb{R}$ be continuous functions that are differentiable on (a, b), and suppose that $g'(s) \neq 0$ for every $s \in (a, b)$. If $f'(s)/g'(s)$ is increasing/decreasing on (a, b), then the function*

$$\phi(s) = \frac{f(s) - f(a)}{g(s) - g(a)}$$

is increasing/decreasing on (a, b).

Next we state an estimate on the mean-squared displacement of the process X, which is the (space) discrete analog of Lemma 4.3.2.

LEMMA 4.3.10. *Suppose Assumption 4.1 holds. Then there exist $h_c > 0$ and $C_1 > 0$ such that the solution of the stochastic equation (4.3.1) satisfies:*

$$\mathbb{E}_x |X(t) - x|^2 \le (C_1 + V(x))\, t\, e^t\,,$$

for any $t \ge 0$ and for all $h < h_c$.

As a consequence, if $t \sim O(\epsilon)$ then the mean-squared displacement of the approximation is $O(\epsilon)$. We emphasize that this lemma is uniform in the spatial step size. This uniformity in h follows from the fact that the stochastic Lyapunov function from Lemma 4.2.6 satisfies (4.2.11) with constants K and γ that do not depend on h.

PROOF. Define $\varphi(y) = (y-x)^T(y-x)$. Since

$$|D\varphi(y)\cdot\eta| \le 2(|y|+|x|)|\eta| \quad \text{for every } x,y,\eta\in\mathbb{R}^n$$

it follows that φ is locally Lipschitz continuous with a local Lipschitz constant that satisfies the sufficient conditions given in Lemma 4.3.8. Thus, Dynkin's formula holds for φ. Also, recall that μ^h denotes the drift field associated to the stochastic equation (4.1.12).

By Dynkin's formula and (4.1.8), we obtain the bound:

$$\begin{aligned}
\mathbb{E}_x\varphi(X(t)) &= \varphi(x) + \mathbb{E}_x\int_0^t Q\varphi(X(s))\,ds \\
&= \mathbb{E}_x\int_0^t 2(X(s)-x)^T\mu^h(X(s))\,ds + \mathbb{E}_x\int_0^t \sum_{i=1}^n 2\cosh\left(\mu_i(X(s))\frac{h}{2}\right)\,ds \\
&\le \mathbb{E}_x\int_0^t \varphi(X(s))\,ds + \mathbb{E}_x\int_0^t |\mu^h(X(s))|^2\,ds \\
&\quad + \mathbb{E}_x\int_0^t \sum_{i=1}^n 2\cosh\left(\mu_i(X(s))\frac{h}{2}\right)\,ds
\end{aligned}$$

Here we used the elementary inequality $2x^Ty \le |x|^2+|y|^2$. To bound the term $(\mu_i^h)^2$, apply Lemma 4.3.9 with $f(s) = (2\sinh(s\mu_i/2))^2$ and $g(s)=s^2$ to obtain

$$\frac{f'(s)}{g'(s)} = \frac{2\mu_i\cosh\left(\mu_i\frac{s}{2}\right)\sinh\left(\mu_i\frac{s}{2}\right)}{s}.$$

Apply Lemma 4.3.9 once more to obtain

$$\frac{f''(s)}{g''(s)} = \mu_i^2\left(\cosh^2\left(\mu_i\frac{s}{2}\right) + \sinh^2\left(\mu_i\frac{s}{2}\right)\right)$$

which is monotonically increasing with respect to s. Hence, there exists an $h_c>0$ such that:

$$\lim_{s\to 0}(\mu_i^s)^2 = \mu_i^2 \le (\mu_i^h)^2 \le \frac{4}{h_c^2}\left(\sinh\left(\mu_i\frac{h_c}{2}\right)\right)^2$$

for all $h<h_c$ and for any $1\le i\le n$. This inequality enables us to bound $(\mu_i^h)^2$ uniformly with respect to h. Owing to the form of the stochastic Lyapunov function in (4.2.10), there exist positive constants C_0 and C_1 such that

$$\begin{aligned}
\mathbb{E}_x\varphi(X(t)) &\le \int_0^t \mathbb{E}_x\varphi(X(s))\,ds + \int_0^t (C_0 + \mathbb{E}_xV(X(s)))\,ds \\
&\le \int_0^t \mathbb{E}_x\varphi(X(s))\,ds + (C_1+V(x))\,t
\end{aligned}$$

To be sure, here we used the maximal polynomial growth condition on the drift field from Assumption 4.1 (A2) and the stochastic Lyapunov function V of the approximation from Lemma 4.2.6. Apply Gronwall's Lemma to this last inequality to finish the proof. $\square$

The next lemma quantifies the complexity of the approximation.

LEMMA 4.3.11. *Suppose Assumption 4.1 holds on the drift field. Let $\bar{N}(t)$ denote the average number of jumps of the approximation $X(t)$ on the interval $[0, t]$. Then there exist positive constants h_c, C_1 and C_2 such that*

$$\bar{N}(t) \leq \frac{C_1\, t}{h^2}(V(x) + C_2)$$

for all $t \geq 0$ and $h < h_c$.

This lemma is not a surprising result: it states that the average number of steps of the SSA method in Algorithm 2.1 is about $\lfloor t/\bar{\delta t} \rfloor$, where t is the time span of simulation and $\bar{\delta t}$ is the mean holding time of the approximation, which from (4.1.8) is proportional to h^2. Since each step requires a function evaluation, this lemma provides a crude estimate for the cost of the approximation as $h \to 0$. Figures 3.8 and 3.30 verify this estimate in the context of a log-normal process that uses adaptive mesh refinement, and a 39-dimensional Brownian Dynamics simulation, respectively.

PROOF. Let $\{t_i\}_{i=0}^{\infty}$ be a sequence of random jump times for the process X with $t_0 = 0$. The time spent by X in $X(t_i)$ (or $t_{i+1} - t_i$) is an exponentially distributed random variable with parameter $\lambda(X(t_i))$, where λ is defined as

$$\lambda(x) = \frac{2}{h^2} \sum_{i=1}^{n} \cosh\left(\mu_i(x)\frac{h}{2}\right).$$

On each time interval (t_i, t_{i+1}) the process is constant, which implies that

$$\int_0^t \lambda(X(s))ds = \sum_{i=0}^{\infty} \lambda(X(t_i))(t \wedge t_{i+1} - t \wedge t_i).$$

Since the drift field μ satisfies a maximal polynomial growth condition from Assumption 4.1 (A2), it is dominated by the stochastic Lyapunov function $V(x)$ from Lemma 4.2.6, and hence,

$$\mathbb{E}_x \int_0^t \lambda(X(s))ds \leq \frac{2}{h^2} \int_0^t \mathbb{E}_x V(X(s))ds \leq \frac{2\,t}{h^2}\,(V(x) + \frac{K}{\gamma}) \tag{4.3.4}$$

where γ and K are taken from Lemma 4.2.6. Thus, we can invoke Fubini's theorem to interchange expectation (conditional on $X(0) = x$) and summation to get:

$$\mathbb{E}_x \int_0^t \lambda(X(s))ds = \sum_{i=0}^{\infty} \mathbb{E}_x \lambda(X(t_i))(t \wedge t_{i+1} - t \wedge t_i)\,,$$

Since conditional on $X(t_i)$ the random variable $(t \wedge t_{i+1} - t \wedge t_i)$ is independent of $\lambda(X(t_i))$, we have that

$$\begin{aligned}
\mathbb{E}_x \int_0^t \lambda(X(s))ds &= \sum_{i=0}^{\infty} \mathbb{E}_x \left(\mathbb{E}_x(\lambda(X(t_i))(t \wedge t_{i+1} - t \wedge t_i) \mid \mathcal{F}_{t_i})\right) \\
&= \sum_{i=0}^{\infty} \mathbb{E}_x(\lambda(X(t_i))\lambda(X(t_i))^{-1}\chi_{t_i \leq t}) \\
&= \bar{N}(t).
\end{aligned}$$

Here χ is an indicator function, $\mathcal{F}_t$ is the natural filtration of the process $X(t)$, and we used the law of iterated expectation to write the total expectation up to time t_{i+1} in terms of the conditional expectation up to time $t = t_i$ using $\mathcal{F}_t$. The desired estimate is obtained by combining the last identity with the inequality in (4.3.4). □

4.4. Generator Accuracy

The following lemma improves upon Prop. 2.5.1 by giving a remainder term estimate.

LEMMA 4.4.1. *Suppose Assumption 4.1 holds. Let $f \in C^4(\mathbb{R}^n)$, and for every $x \in \mathbb{R}^n$, define $Rf(x)$ as*

$$Rf(x) = Lf(x) - Qf(x).$$

Then for all $x \in \mathbb{R}^n$, there exists $C_f(x) > 0$ such that:

$$|Rf(x)| \leq h^2 C_f(x).$$

PROOF. Retracing (and strengthening) the proof of Prop. 2.5.1 using Taylor's integral formula (in place of Maclaurin expansions) yields:

$$Qf(x) = Lf(x) + \underbrace{h^2 \sum_{i=1}^{n} (A_i + B_i + C_i + D_i^+ + D_i^-)}_{=-Rf(x)}, \tag{4.4.1}$$

where we have introduced:

$$\begin{aligned}
A_i &= \frac{\mu_i^3}{8} \frac{\partial f}{\partial x_i} \int_0^1 \cosh\left(s\mu_i \frac{h}{2}\right) (1-s)^2 ds \\
B_i &= \frac{\mu_i^2}{4} \frac{\partial^2 f}{\partial x_i^2} \int_0^1 \cosh\left(s\mu_i \frac{h}{2}\right) (1-s) ds \\
C_i &= \frac{\mu_i}{6} \frac{\partial^3 f}{\partial x_i^3} \int_0^1 \cosh\left(s\mu_i \frac{h}{2}\right) ds \\
D_i^{\pm} &= \frac{1}{6} \exp\left(\pm \mu_i \frac{h}{2}\right) \int_0^1 \frac{\partial^4 f}{\partial x_i^4}(x \pm s e_i h)(1-s)^3 ds
\end{aligned}$$

Applying the elementary inequality:

$$\cosh(sw) \leq \cosh(w) \quad \text{for any } s \in [0,1] \text{ and } w \in \mathbb{R}$$

and simplifying yields the desired estimate with

$$\begin{aligned}
C_f(x) &= C \sum_{i=1}^{n} \cosh\left(\mu_i \frac{h}{2}\right) \\
&\times \left(|\mu_i|^3 \left| \frac{\partial f}{\partial x_i} \right| + |\mu_i|^2 \left| \frac{\partial^2 f}{\partial x_i^2} \right| + |\mu_i| \left| \frac{\partial^3 f}{\partial x_i^3} \right| + \sup_{|s|<1} \left| \frac{\partial^4 f}{\partial x_i^4}(x + s e_i h) \right| \right)
\end{aligned} \tag{4.4.2}$$

where C is a positive constant. □

4.5. Global Error Analysis

In this section it is shown that the approximation associated to the generator Q in (4.1.8) is weakly accurate on finite time intervals. Moreover, it is shown that the stationary probability density of Q is nearby the true one. These statements are made in the context of the model SDE (4.1.5) under Assumption 4.1. The proof is a special case of a general framework for quantifying the rate of convergence of numerical solutions of diffusions (even controlled ones) introduced in [**34**]. This approach assumes sufficient regularity in the statistic to be approximated, and requires a stochastic representation of both the desired statistic and its approximation in terms of the approximating process. The regularity of the statistic to be approximated is then exploited to obtain an expansion of the error as a function of the discretization parameter.

4.5.1. Accuracy Theorems. We are now in position to estimate the global error of the numerical method.

THEOREM 4.5.1 (Finite-Time Accuracy). *Suppose Assumption* 4.1 *holds. For every* $\varphi \in C_b^4(\mathbb{R}^n)$*, for any* $t \geq 0$*, and for all* $x \in S_h$*, there exists a positive constant* $C_{\varphi,t}$ *such that*

$$|P_t^h \varphi(x) - P_t \varphi(x)| \leq h^2 \, C_{\varphi,t} \, V(x). \tag{4.5.1}$$

The main issue in the subsequent proof is that the generators of the SDE solution and numerical approximation may be unbounded, since the drift field entering (4.1.5) and (4.1.12) satisfies the minimal growth condition Assumption 4.1 (A4). We get around this possible issue by using: (i) regularity of the quantity being approximated; (ii) a stochastic representation of the global error in terms of the approximating process; and (iii) a stochastic Lyapunov function for the approximation from Lemma 4.2.10.

PROOF. Define the global error of the approximation as:

$$\epsilon(x,t) = P_t \varphi(x) - P_t^h \varphi(x) \quad \text{for any } x \in S_h \text{ and } t \geq 0. \tag{4.5.2}$$

Following the method of proof given in [**34**], we find a stochastic representation for this error in terms of the function being approximated and the approximation itself. We start with a derivation that assumes the generators are better behaved than they actually are, and then later we justify this formula.

If the generators were bounded with respect to bounded functions, then the global error satisfies an inhomogeneous linear differential equation:

$$\dot{\epsilon}(x,t) - Q\epsilon(x,t) = (L - Q) P_t \varphi(x) \quad \text{for any } x \in S_h \text{ and } t \geq 0 \tag{4.5.3}$$

with initial condition:

$$\epsilon(x, 0) = 0 \quad \text{for all } x \in S_h.$$

A variation of constants (or Duhamel) formula then suggests that the global error of the approximation can be written as:

$$\epsilon(x,t) = \int_0^t P_{t-s}^h (L - Q) P_s \varphi(x) ds \tag{4.5.4}$$

which has the following stochastic representation:

$$P_t \varphi(x) - P_t^h \varphi(x) = \int_0^t \mathbb{E}_x \{ (L - Q) P_r \varphi(X(t - r)) \} dr.$$

Note that the expectation in the integrand is over the approximating process $X(t-r)$.

In order to justify this formula, it is convenient to work with the transition evolution operators of, respectively, the SDE solution and the approximation:

$$\begin{cases} P_{s,t}\varphi(x) = \mathbb{E}\{\varphi(Y(t)) \mid \varphi(Y(s)) = x\} \\ P^h_{s,t}\varphi(x) = \mathbb{E}\{\varphi(X(t)) \mid \varphi(X(s)) = x\} \end{cases}$$

and since both Y and X are time-homogeneous, these evolution operators satisfy:

$$P_{s,t} = P_{t-s} \quad \text{and} \quad P^h_{s,t} = P^h_{t-s}$$

for any $t \geq s \geq 0$. An application of Itō's backward formula for diffusion processes shows that:

$$P_{s,t}\varphi(x) = \varphi(x) - \int_s^t LP_{r,t}\varphi(x)dr$$

An application of Itō's formula for semimartingales implies that:

$$P^h_{s,t}\varphi(x) = \varphi(x) + \int_s^t P^h_{s,r}Q\varphi(x)dr$$

Combining these results gives

$$P^h_{t,s}P_{s,t}\varphi(x) = \varphi(x) + \int_t^s P^h_{t,r}(Q-L)P_{r,t}\varphi(x)dr$$

and hence,

$$P_{s,t}\varphi(x) = P^h_{s,t}\varphi(x) + \int_t^s P^h_{s,r}(Q-L)P_{r,t}\varphi(x)dr.$$

Setting $s = 0$, writing this expression in terms of the semigroups corresponding to Q and L, and simplifying yields

$$P_t\varphi(x) - P^h_t\varphi(x) = \int_0^t P^h_{t-r}(L-Q)P_r\varphi(x)dr$$

which is the variations of constants formula given in (4.5.4) for the global error of the approximation in terms of the local error of Q with respect to the true solution.

In order to estimate this global error, it helps to express this formula in terms of transition probabilities as:

$$\epsilon(x,t) = \int_0^t \sum_{z\in S_h} \Pi^h_{t-s,x}(z)(L-Q)P_s\varphi(z)ds.$$

Apply Lemma 4.4.1 to obtain

$$\begin{aligned} |\epsilon(x,t)| &= |\int_0^t \sum_{z\in S_h} \Pi^h_{t-s,x}(z)(L-Q)P_s\varphi(z)ds| \\ &\leq \int_0^t \sum_{z\in S_h} \Pi^h_{t-s,x}(z)|(L-Q)P_s\varphi(z)|ds \\ &\leq h^2 \int_0^t \sum_{z\in S_h} \Pi^h_{t-s,x}(z)C_{P_s\varphi}(z)ds \end{aligned}$$

where $C_{P_s\varphi}(x)$ is given in (4.4.2) with $f = P_s\varphi$. Note that there exists $C_0(t) > 0$ such that,

$$C_{P_s\varphi}(x) < V(x) + C_0(t)$$

for all $x \in \mathbb{R}^n$ and for all $s \in [0,t]$. Indeed, from Lemma 4.3.6 if $\varphi \in C_b^4(\mathbb{R}^n)$ then $P_s\varphi \in C_b^4(\mathbb{R}^n)$ for any $s \geq 0$, and by Assumption 4.1 (A2) on the drift field the terms that involve the drift field in (4.4.2) are dominated by the Lyapunov function $V(x)$. Hence, we obtain

$$\begin{aligned}|\epsilon(x,t)| &\leq h^2 \int_0^t \sum_{z\in S_h} \Pi^h_{t-s,x}(z)(V(z) + C_0(t))ds \\ &\leq t\, h^2 \,(V(x) + \tilde{C}_0(t))\end{aligned}$$

for some $\tilde{C}_0(t) > 0$. □

In addition, the Markov process generated by Q accurately represents the stationary distribution of the SDE.

THEOREM 4.5.2 (Stationary Distribution Accuracy). *Suppose Assumption* 4.1 *holds. For every $\varphi \in C_b^4(\mathbb{R}^n)$, we have that*

$$|\Pi^h(\varphi|_{S_h}) - \Pi(\varphi)| \leq h^2\, C_\varphi \tag{4.5.5}$$

where

$$C_\varphi = \sup_{0<h\leq h_c} \Pi^h\left(\int_0^\infty C_{P_s\varphi} ds\right)$$

and $C_{P_s\varphi}(x)$ is given in (4.4.2) *with $f = P_s\varphi$.*

PROOF. Note that this result is not a corollary of Theorem 4.5.1, since the latter gives an upper bound on the global error that is increasing with t. In order to obtain the desired estimate, we use a slightly different approach that begins with averaging the global error with respect to the invariant probability measure of the approximation Π^h (which exists by Theorem 4.2.8):

$$\begin{aligned}|\Pi^h(\epsilon(t))| &= |\sum_{z\in S_h} \Pi^h(z)\epsilon(t,z)| \\ &= |\Pi^h\left(\int_0^t (L-Q)P_s\varphi ds\right)| \\ &\leq \Pi^h\left(\int_0^t |(L-Q)P_s\varphi| ds\right) \\ &\leq h^2 \Pi^h\left(\int_0^t C_{P_s\varphi} ds\right)\end{aligned}$$

where $C_{P_s\varphi}(x)$ is given in (4.4.2) with $f = P_s\varphi$. Here we used: (i) the Fubini Theorem to interchange space and time integrals; (ii) the remainder term estimate from Lemma 4.4.1; and (iii) the fact that $\Pi^h(P_t^h\varphi) = \Pi^h(\varphi)$. Passing to the limit as $t \to \infty$, and using Theorem 4.2.5, gives the desired estimate. □

REMARK 4.5.1. One may be able to prove Theorem 4.5.2 by adapting to this setting the version of Stein's method developed in §6.2 of [**117**].

REMARK 4.5.2. In the context of smooth initial conditions and a smooth drift field with at most polynomial growth at infinity, an estimate of the type:

$$\|D^j P_t \varphi(x)\| \le C(1+|x|^s)e^{-\gamma t}\,, \quad \forall\, t \ge 0\,, \quad \forall\, j \ge 1\,, \quad \forall\, x \in \mathbb{R}^n\,,$$

is available in Theorem 3.1 of [**152**]. Roughly speaking, the proof of this theorem shows exponential convergence of the first n derivatives of the SDE semigroup P_t in $L^2(\mathbb{R}^n)$ and then uses the Sobolev Embedding Theorem to transfer these estimates to $L^\infty(\mathbb{R}^n)$, i.e., trade regularity for integrability. This procedure leads to point wise estimates on the derivatives of the semigroup. One may be able to derive similar estimates in this context by assuming that the first n derivatives of the drift field exist, but we wish to avoid making this assumption.

CHAPTER 5

Analysis on Gridless State Spaces

To study issues that may arise when the noise is multiplicative (or state-dependent) or if the spatial step size is adaptive, consider the generator in (4.1.8) but, with random spatial step size equal to ξh where $\xi \in U(1/2,1)$. Figure 5.1 illustrates the difference between points spaced in this variable way and equally spaced points on a grid. In particular, with random spacing the state space of the approximation becomes gridless. In this chapter we adapt the stability/accuracy results from Chapter 4 to this setting. See §2.2 for the definition of a gridless state space.

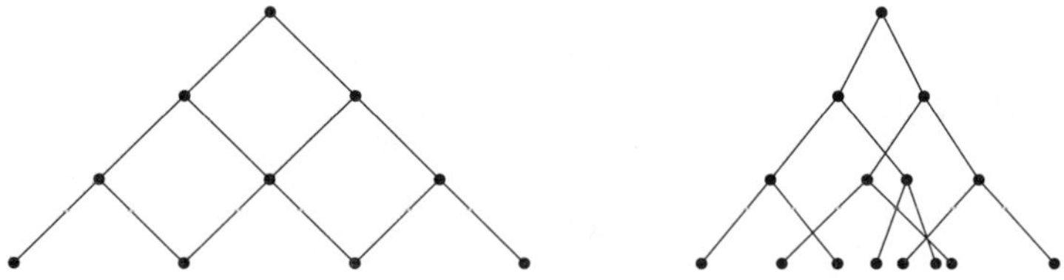

(a) grid (b) gridless

FIGURE 5.1. **Gridded vs Gridless State Spaces.** This figure illustrates the difference between gridded and gridless state spaces. Points in these spaces are produced from a given seed in $\mathbb{R}$ using fixed spatial step size h (left panel) and random spatial step size ξh (right panel) where $\xi \sim U(1/2,1)$. The set of all such states is the state space of the Markov jump process. In the case of a gridded state space, every pair of points is connected by a finite path on the grid. Thus, if the jump rates are positive at every point, then the Markov process is irreducible on this gridded state space. In contrast, in the gridless case, there exist pairs of nodes such that no finite path on the graph has those nodes as endpoints. Thus, even if the jump rates are positive at every point, the Markov process is not irreducible on this collection of points.

5.1. A Random Walk in a Random Environment

Here we introduce a realizable spatial discretization that uses variable step sizes. To introduce this discretization, in addition to the notation presented in Chapter 4: let ξ to be a Gaussian-related random field defined as the map $x \mapsto g(\tilde{\xi}(x))$ where $\tilde{\xi} : \mathbb{R}^n \to \mathbb{R}$ is a smooth stationary Gaussian random field and $g : \mathbb{R} \to [1/2,1]$ is the smooth, invertible function $g(x) = \arctan(x)/(2\pi) + 3/4$. We stress that

$$h_{\min} \le \xi(x) h \le h_{\max} \quad \text{for all } x \in \mathbb{R}^n$$

where $h_{\min} = h/2$ and $h_{\max} = h$. We stress that the statements in this chapter are conditional on ξ. In particular, we do not average over the random field ξ.

Even though the approximation uses variable step sizes, its state space is still countable. Indeed, from a given state only finitely many states can be reached by the approximation in a finite number of steps, and hence, only countably many states can be reached in total, which implies that the state space of the approximation is countable. However, the approximation is not irreducible on this gridless state space as illustrated in Figure 5.1. In addition, if we try to embed the state space of the approximation into $\mathbb{R}^n$, the process is still not irreducible with respect to the standard topology on $\mathbb{R}^n$, since at any finite time there may be zero probability for the process to reach an open neighborhood of a designated point in state space starting from any other point. To deal with this lack of irreducibility, we mollify the generator (5.1.1) so that Q_q is a bounded linear operator and its associated semigroup satisfies the Feller property. This additional regularity on Q_q is sufficient to generalize many of the results from Chapter 4 to an approximation with a gridless state space.

Given $r \in \mathbb{N}_0$, the random field ξ, a spatial step size parameter h, and a test function $f : \mathbb{R}^n \to \mathbb{R}$, define the infinitesimal generator of the approximation on a gridless state space as follows.

$$\boxed{\begin{aligned} Q_q f(x) = {} & \frac{\exp\left(-\,\xi(x)^2 h^2\, |x|^{2r}\right)}{\xi(x)^2 h^2} \\ & \times \left(\sum_{i=1}^{n} \left(\exp\left(+\xi(x) h \frac{\mu_i(x)}{2} \right) (f(x + \xi(x) h e_i) - f(x)) \right.\right. \\ & \left.\left. + \exp\left(-\xi(x) h \frac{\mu_i(x)}{2} \right) (f(x - \xi(x) h e_i) - f(x)) \right) \right) \end{aligned}} \tag{5.1.1}$$

This generator induces an approximation on a gridless state space, which is a continuous-time random walk in a random environment [**147**, **163**]. The prefactor in (5.1.1) will ensure that the jump rates entering Q_q are bounded, which implies that Q_q is a bounded linear operator. We derive conditions on the natural number r which also imply that Q_q has a stochastic Lyapunov function.

5.2. Generator Accuracy

Here we state that Q_q is accurate in approximating the infinitesimal generator of the SDE.

LEMMA 5.2.1. *Suppose Assumption* 4.1 *holds. Let* $f \in C^4(\mathbb{R}^n)$ *and set*

$$R_q f(x) = L f(x) - Q_q f(x).$$

For every $x \in \mathbb{R}^n$*, there exists* $C_f(x) > 0$ *(uniform in* ξ*) such that:*

$$|R_q f(x)| \le h^2\, C_f(x).$$

PROOF. This result follows immediately from the proof of Lemma 4.4.1. Specifically, repeat the steps with h replaced by ξh without expanding the mollifier, and as a last step, bound the mollifier in $Q_q f(x)$ by unity and use the fact that $\xi h < h$ to obtain:

$$|R_q f(x)| \le \xi^2 h^2 \tilde{C}_f(x) \le h^2 C_f(x)$$

Note that the positive function $C_f(x)$ in this upper bound is the one appearing in Lemma 4.4.1, i.e.,

$$\begin{aligned} C_f(x) &= C \sum_{i=1}^{n} \cosh\left(\mu_i \frac{h}{2}\right) \\ &\times \left(|\mu_i|^3 \left|\frac{\partial f}{\partial x_i}\right| + |\mu_i|^2 \left|\frac{\partial^2 f}{\partial x_i^2}\right| + |\mu_i| \left|\frac{\partial^3 f}{\partial x_i^3}\right| + \sup_{|s|<1} \left|\frac{\partial^4 f}{\partial x_i^4}\right| (x + s e_i h)\right) \end{aligned} \tag{5.2.1}$$

where C is a positive constant. □

5.3. Stability by Stochastic Lyapunov Function

Here we show that the approximation on a gridless state space has a stochastic Lyapunov function.

LEMMA 5.3.1. *Suppose the drift field of the SDE* (4.1.5) *satisfies Assumption* 4.1. *Let*

$$V(x) = \exp\left(a|x|_{2m+2}^{2m+2}\right) \quad \textit{for any } 0 < a < \frac{\beta_0}{2(m+1)} \tag{5.3.1}$$

where m and β_0 are the constants appearing in Assumption 4.1 *(A2). Then, there exist positive constants K and γ (uniform in h and ξ) such that*

$$Q_q V(x) \le K - \gamma V(x) \tag{5.3.2}$$

holds for all $x \in \mathbb{R}^n$, h sufficiently small, and $r < 2m+1$.

PROOF. The steps in this proof are similar to the proof of Lemma 4.2.6, but there are some differences which we highlight below. In particular, the proof shows that there exist h_c, $R^\star$ and $\gamma > 0$ such that:

$$\frac{Q_q V(x)}{V(x)} < -\gamma \quad \text{for all } x \in \mathbb{R}^n \text{ satisfying } |x| > R^\star \tag{5.3.3}$$

and for all $h < h_c$. We then choose K in (5.3.2) to satisfy:

$$K \ge \sup_{|x| \le R^\star} |LV(x) + \gamma V(x)| + \sup_{0<h<h_c} \sup_{|x|\le R^\star} |(Q_q - L)V(x)|.$$

The first two terms of K are independent of h and $\xi(x)$ since $V(x)$ is smooth and the set $\{|x| \le R^\star\}$ is compact. According to Lemma 5.2.1, the difference $|(Q_q - L)V(x)|$ is $\mathcal{O}(h^2)$, involves four derivatives of $V(x)$, and is independent of the random environment. Since $V(x)$ is smooth and the set $\{|x| \le R^\star\}$ is compact, the last term can be bounded uniformly in h. Thus, we obtain the desired estimate if we can find $R^\star$ (uniform in h and ξ) such that (5.3.3) holds.

Following (Step 1) and (Step 2) of the proof of Lemma 4.2.6, we apply the generator Q_q in (5.1.1) to $V(x)$ to obtain:

$$\frac{Q_q V(x)}{V(x)} = \sum_{i=1}^{n} A_i(x) \exp\left(-\xi(x)^2 \, h^2 \, |x|^{2r}\right) \tag{5.3.4}$$

where we have defined:

$$\begin{aligned} A_i(x) &= \frac{1}{\xi^2 h^2} \exp\left(\frac{\xi h}{2}\mu_i\right) \left(\exp\left(a((x_i + \xi h)^2)^{m+1} - a((x_i)^2)^{m+1}\right) - 1\right) \\ &+ \frac{1}{\xi^2 h^2} \exp\left(-\frac{\xi h}{2}\mu_i\right) \left(\exp\left(a((x_i - \xi h)^2)^{m+1} - a((x_i)^2)^{m+1}\right) - 1\right) \end{aligned}$$

Note that $A_i(x)$ involves positive and negative terms, and in order to obtain the estimate (5.3.3) we must show that the negative terms dominate in a way that is uniform in h and ξ. For any $x \in \mathbb{R}^n$, note that Lemma 4.2.7 implies that A_i satisfies

$$\begin{aligned} A_i(x) \le & \frac{1}{\xi^2 h^2} \exp\left(\frac{\xi h}{2}\mu_i\right)\left(\exp\left(a(2x_i\xi h + \xi^2 h^2)(m+1)(x_i+\xi h)^{2m}\right) - 1\right) \\ & + \frac{1}{\xi^2 h^2} \exp\left(-\frac{\xi h}{2}\mu_i\right)\left(\exp\left(a(-2x_i\xi h + \xi^2 h^2)(m+1)(x_i-\xi h)^{2m}\right) - 1\right) \end{aligned}$$

for $1 \le i \le n$.

These bounds on A_i imply that:

$$\frac{Q_q V(x)}{V(x)} \le \phi(h,x) = \exp\left(-\xi^2 h^2 d(x)\right) \times \left(\sum_{i=1}^n \phi_i(h,x)\right) \tag{5.3.5}$$

where we have introduced:

$$\phi_i(s,x) = \frac{2}{\xi^2 s^2}\Big(\exp(c_i(x)\xi^2 s^2)\cosh(a_i(x)\xi s) - \cosh(b_i(x)\xi s)\Big) \tag{5.3.6}$$

and the following functions of position only:

$$\begin{cases} a_i(x) = 2\,a\,(m+1)\,x_i^{2m+1} + \frac{1}{2}\mu_i \\ b_i(x) = \frac{1}{2}\mu_i \\ c_i(x) = C_1 x_i^{2m} + C_0 \\ d(x) = |x|^{2r} \end{cases} \tag{5.3.7}$$

where C_1 and C_0 are positive constants. Comparing (4.2.16) to (5.3.5), note that the main difference between these bounds is the prefactor entering the upper bound in (5.3.5), which comes from the mollification we introduced in Q_q. However, since this prefactor multiplies all of the terms in $\phi(h,x)$, the order of magnitude of the terms in $\phi(h,x)$ is the same as described in (Step 3) of the proof of Lemma 4.2.6. Thus, for any $s > 0$, $\phi(s,x)$ tends to zero from below as $|x| \to \infty$. We underscore that the source of dissipation in $Q_q V(x)$ is the same as in $QV(x)$.

Following (Step 4) of the proof of Lemma 4.2.6, we now derive an upper bound on the right-hand-side of (5.3.5) from the limit of $\phi(h,x)$ as $h \to 0$. By l'Hôpital's Rule this limit is given by:

$$\begin{aligned} \lim_{s\to 0}\phi(s,x) &= \sum_{i=1}^n (a_i^2 - b_i^2) + 2c_i \\ &\le 2a(m+1)(2a(m+1)-\beta)|x|_{4m+2}^{4m+2} + \text{lower order in } |x| \text{ terms} \end{aligned} \tag{5.3.8}$$

Note that if $|x|$ is large enough, $\lim_{s\to 0}\phi(s,x)$ is large and negative. A second application of l'Hôpital's Rule shows that

$$\lim_{s\to 0}\frac{\partial\phi}{\partial s}(s,x) = 0 \tag{5.3.9}$$

and hence, the point $s = 0$ is a critical point of $\phi(s,x)$. To determine the properties of this critical point, we apply l'Hôpital's Rule a third time to obtain:

$$\lim_{s\to 0}\frac{\partial^2\phi}{\partial s^2}(s,x) = \sum_{i=1}^n \frac{\xi^2}{6}(a_i^4 - b_i^4 + 12a_i^2 c_i + 12c_i^2 + 12(b_i^2 - a_i^2 - 2c)d)\xi^2. \tag{5.3.10}$$

It follows from (Step 4) of the proof of Lemma 4.2.6 that the term $a_i^4 - b_i^4$ is negative and satisfies $|a_i^4 - b_i^4| = O(|x|^{8m+4})$. Whereas our preceding analysis showed that the term $(b_i^2 - a_i^2 - 2c)d$ is positive, and it is straightforward to verify that this term satisfies $(b_i^2 - a_i^2 - 2c)d = O(|x|^{2r+4m+2})$. Thus, since ξ is uniformly bounded from below, if $r < 2m+1$ then the negative term dominates and $\lim_{s\to 0} \frac{\partial^2 \phi}{\partial s^2}(s, x)$ is negative if $|x|$ is large enough. By the point-wise second derivative rule, $s = 0$ is a local maximum of the function $\phi(s, x)$, and therefore, we can find h_c such that: for any $h < h_c$, $\phi(h, x)$ is bounded by $\lim_{h\to 0} \phi(h, x) < 0$, as required by the estimate in (5.3.3). Thus, the above bound on $\phi(h, x)$ holds for all $h < h_c$, for all $|x|$ sufficiently large and for any $r < 2m + 1$. □

REMARK 5.3.1. Lemma 5.3.1 indicates that the largest r can be for the approximation to have a stochastic Lyapunov function is $2m+1$. Because the mollification in Q_q is artificial, we would like to pick r to be as small as possible. To globally bound the jump rates in Q_q, the minimal choice of r in (5.1.1) turns out to be $m + 1$. In fact, we show in Lemma 5.5.2 below, this choice is the minimal number such that Q_q is bounded. Moreover, by Lemma 5.3.1, this choice ensures that the approximation has a stochastic Lyapunov function.

Next we apply this stochastic Lyapunov function to study a martingale problem associated to the Markov process induced by Q_q.

LEMMA 5.3.2. *Suppose Assumption* 4.1 *holds on the drift field, let $V(x)$ be the Lyapunov function from Lemma* 5.3.1, *and let $\varphi : \mathbb{R}^n \to \mathbb{R}$. Consider the local martingale:*

$$M^{\varphi}(t) = \varphi(X(t)) - \varphi(X(0)) - \int_0^t Q_q\varphi(X(s))ds, \quad t \geq 0.$$

If φ satisfies the growth condition:

$$Q_q\varphi^2(x) - 2\varphi(x)Q_q\varphi(x) \leq V(x) + C \tag{5.3.11}$$

for some positive constant C and for all $x \in \mathbb{R}^n$, then for any $r < 2m+1$, the local martingale $M^{\varphi}(t)$ is a global martingale.

The proof of Lemma 4.3.7 can be adapted to prove Lemma 5.3.2. It follows from Lemma 5.3.2 that Dynkin's formula (4.1.10) holds for any function satisfying the growth condition (5.3.11). Lemma 4.3.8 (with Q replaced by Q_q) provides a practical way to test if a function satisfies this condition.

5.4. Finite-Time Accuracy

Here we consider finite-time accuracy of the approximation.

THEOREM 5.4.1 (Finite-Time Accuracy). *Suppose Assumption* 4.1 *holds. For every $\varphi \in C_b^4(\mathbb{R}^n)$, for any $t \geq 0$, for all $x \in S_h$, and for any $r < 2m + 1$, there exists a positive constant $C_{\varphi,t}$ such that*

$$|P_t^h\varphi(x) - P_t\varphi(x)| \leq h^2\, C_{\varphi,t}\, V(x). \tag{5.4.1}$$

PROOF. Like the proof of Theorem 4.5.1, the main ingredients in this proof are stability of the approximation (in the sense that it has a stochastic Lyapunov function) and sufficiently regularity in the solution being approximated, which come from Lemmas 5.3.1 and Lemma 4.3.6 respectively. With these lemmas in hand, the desired error estimate can be obtained by repeating the steps used in the proof of Theorem 4.5.1. □

5.5. Feller Property

Here we prove a regularity property of the semigroup associated to the infinitesimal generator Q_q. As we show below, this regularity combined with the stochastic Lyapunov function from Lemma 5.3.1 imply existence of an invariant probability measure for the approximation.

LEMMA 5.5.1. *Suppose Assumption* 4.1 *holds. If the natural number* r *in* (5.1.1) *satisfies* $r \geq m+1$, *then for any* $h > 0$ *the linear operator* Q_q *is bounded.*

PROOF. It suffices to show that there exists $C > 0$ such that

$$\|Q_q f\|_\infty \leq C \, \|f\|_\infty \, , \tag{5.5.1}$$

for an arbitrary $f \in B_b(\mathbb{R}^n)$. In order to prove this statement, let $J_i^{\pm}(x)$ be the jump rate from x to the state $y_i^{\pm}(x) = x \pm e_i \xi(x) h$. By Assumption 4.1 (A2) the drift field satisfies a maximal polynomial growth condition with parameter C_P^{μ}, and by our hypothesis $r \geq m+1$, we have that

$$\begin{aligned} J_i^{\pm}(x) &= \frac{1}{\xi(x)^2 h^2} \exp\left(\pm \frac{h\xi(x)}{2} \mu_i(x) - \xi(x)^2 h^2 \, |x|^{2r} \right) \\ &\leq \frac{1}{\xi(x)^2 h^2} \exp\left(h C_P^{\mu} \left(1 + |x|^{2m+1}\right) - \xi(x)^2 h^2 \, |x|^{2r} \right) \\ &\leq C(h) \end{aligned}$$

for all $x \in \mathbb{R}^n$ and for any $1 \leq i \leq n$. Apply this bound on the jump rates to obtain the following bound on $Q_q f$,

$$\begin{aligned} \|Q_q f\|_\infty &= \sup_{x \in \mathbb{R}^n} \left| \sum_{i=1}^{n} J_i^{+}(x)(f(y_i^{+}) - f(x)) + J_i^{-}(x)(f(y_i^{-}) - f(x)) \right| \\ &\leq 2\|f\|_\infty \sup_{x \in \mathbb{R}^n} \sum_{i=1}^{n} (J_i^{+}(x) + J_i^{-}(x)) \\ &\leq 4 \, n \, C(h) \, \|f\|_\infty \end{aligned}$$

which shows that the generator Q_q is bounded. □

Since Q_q is a bounded linear operator, the semigroup it induces is Feller, as the next lemma states.

LEMMA 5.5.2. *Suppose Assumption* 4.1 *holds. For any* $t \geq 0$ *and for any* $r \geq m+1$, *the semigroup* P_t^h *generated by* Q_q *is Feller, which means that for each* $f \in C_b(\mathbb{R}^n)$ *we have that* $P_t^h f \in C_b(\mathbb{R}^n)$.

PROOF. Define $P_t^h = \exp(Q_q t)$ by its exponential series expansion. Since Q_q is a bounded linear operator on the Banach space $C_b(\mathbb{R}^n)$, then the operator P_t^h is bounded on $C_b(\mathbb{R}^n)$, and the homogeneous Cauchy problem associated to Q_q has a unique classical solution for every initial condition in $C_b(\mathbb{R}^n)$. Moreover, this solution belongs to $C_b(\mathbb{R}^n)$ for all $t \geq 0$. □

5.6. Stationary Distribution Accuracy

The next result states that the approximation has at least one invariant probability measure.

LEMMA 5.6.1. *Suppose Assumption* 4.1 *holds on the drift field. Then, for any* $m+1 \le r < 2m+1$, *the Markov process generated by* Q_q *has at least one invariant probability measure.*

PROOF. This result follows by the classical Krylov-Bogolyubov argument [**130**], since by Lemma 5.5.2, the semigroup associated to Q_q is Feller if $r \ge m+1$, and by Lemma 5.3.1, the process generated by Q_q has a stochastic Lyapunov function if $r < 2m+1$. □

With this invariant probability measure in hand, we obtain the following extension of Theorem 4.5.2.

THEOREM 5.6.2 (Stationary Distribution Accuracy). *Suppose Assumption* 4.1 *holds. For every* $\varphi \in C_b^4(\mathbb{R}^n)$ *and for any* $m+1 \le r < 2m+1$, *we have*

$$|\Pi^h(\varphi) - \Pi(\varphi)| \le \; h^2 \, C_\varphi \tag{5.6.1}$$

where C_φ *is defined as:*

$$C_\varphi = \sup_{0<h\le h_c} \Pi^h \left(\int_0^\infty C_{P_s\varphi} ds \right)$$

and $C_{P_s\varphi}(x)$ *is given in* (5.2.1) *with* $f = P_s\varphi$.

The proof of this theorem is identical to the proof of Theorem 5.6.2, and therefore, omitted.

CHAPTER 6

Tridiagonal Case

The infinitesimal generators we constructed in §2.3 for scalar SDEs have an infinite tridiagonal matrix representation. In this Chapter, we use this representation to show that the invariant density, exit probability, and mean first passage time of these approximations satisfy a tridiagonal system of equations, which we explicitly solve. We obtain explicit formulas for the solutions to these equations. These formulas were used to numerically test our approximations on the scalar SDEs presented in §3.2, §3.3 and §3.6. Most of the following results are classical, but their application to numerical methods for SDEs seems somewhat new.

6.1. Invariant Density

Let $S = \{x_i\}_{i \in \mathbb{Z}}$ be a collection of grid points over $\mathbb{R}$ ordered by index so that:

$$\dots < x_{i-1} < x_i < x_{i+1} < \dots$$

The one-dimensional generators in (2.3.7), (2.3.10), and (2.3.16) can all be written as a tridiagonal operator Q whose action on a grid function $f : S \to \mathbb{R}$ is given by:

$$(Qf)_i = Q_{i,i+1} f_{i+1} - (Q_{i,i+1} + Q_{i,i-1}) f_i + Q_{i,i-1} f_{i-1} \tag{6.1.1}$$

where we used the shorthand notation: $f_i = f(x_i)$ for any grid point $x_i \in S$. Let Q^* denote the transpose of Q, which satisfies $(Q^*)_{i,j} = Q_{j,i}$ for all $i, j \in \mathbb{Z}$. Recall, that an invariant (not necessarily integrable) density of Q is a grid function $\nu_d : S \to \mathbb{R}$ that solves:

$$(Q^* \nu_d)_i = Q_{i+1,i} (\nu_d)_{i+1} - (Q_{i,i+1} + Q_{i,i-1}) (\nu_d)_i + Q_{i-1,i} (\nu_d)_{i-1} = 0 \tag{6.1.2}$$

for every $i \in \mathbb{Z}$. A generator of the form (6.1.1) is always 'ν_d-symmetric' as the next Proposition states.

PROPOSITION 6.1.1. *Let $\nu_d : S \to [0, \infty)$ be a grid function that satisfies $(\nu_d)_0 = 1$ and $(\nu_d)_1 = Q_{0,1}/Q_{1,0}$. Then ν_d satisfies* (6.1.2) *if and only if Q is ν_d-symmetric, in the sense that*

$$Q_{i+1,i} (\nu_d)_{i+1} = Q_{i,i+1} (\nu_d)_i \quad \textit{for all } i \in \mathbb{Z}. \tag{6.1.3}$$

PROOF. The ν_d-symmetry condition (6.1.3) immediately implies the invariant density condition (6.1.2). For the converse, use induction: our hypothesis on the seed values implies that (6.1.3) holds for $i = 0$. Assume (6.1.3) holds for $i = j - 1 > 0$. Since ν_d satisfies (6.1.2), it follows that

$$Q_{j+1,j} (\nu_d)_{j+1} - Q_{j,j+1} (\nu_d)_j = Q_{j,j-1} (\nu_d)_j - Q_{j-1,j} (\nu_d)_{j-1}$$

The inductive hypothesis implies that the right-hand-side of this equation is zero, and therefore, (6.1.3) holds for all $i \geq 0$. A similar proof works for $i < 0$ and is therefore omitted. □

In addition to ν_d-symmetry, a closed form solution to the recurrence relation in (6.1.2) is available.

PROPOSITION 6.1.2 (Formula for Invariant Density). *A formula for the solution to* (6.1.2) *with seed values* $(\nu_d)_0 = 1$ *and* $(\nu_d)_1 = Q_{0,1}/Q_{1,0}$ *is given by*

$$(\nu_d)_i = \begin{cases} \displaystyle\prod_{k=0}^{|i|-1} \frac{Q_{k,k+1}}{Q_{k+1,k}} & \text{if } i > 0 \\ \displaystyle\prod_{k=0}^{|i|-1} \frac{Q_{-k,-k-1}}{Q_{-k-1,-k}} & \text{if } i < 0 \\ 1 & \text{if } i = 0 \end{cases} \tag{6.1.4}$$

This expression for ν_d in terms of the entries of Q can be easily implemented in e.g. MATLAB.

PROOF. The proof shows that ν_d in (6.1.4) satisfies ν_d-symmetry, which by Prop. 6.1.1 implies that ν_d solves (6.1.2). Evaluating (6.1.2) at $i = 0$ implies that the formula (6.1.4) holds for $i = \pm 1$, since $(\nu_d)_{\pm 1} = Q_{0,\pm 1}/Q_{\pm 1,0}$. For $i > 1$,

$$(\nu_d)_{i+1} Q_{i+1,i} = \overbrace{\left(\prod_{k=0}^{i} \frac{Q_{k,k+1}}{Q_{k+1,k}} \right)}^{(\nu_d)_{i+1}} Q_{i+1,i} = \underbrace{\left(\prod_{k=0}^{i-1} \frac{Q_{k,k+1}}{Q_{k+1,k}} \right)}_{(\nu_d)_i} Q_{i,i+1} = (\nu_d)_i Q_{i,i+1}$$

which implies the desired result for $i \geq 0$. A similar result holds for $i < -1$. □

A Markov process on a countable state space is ergodic (or positive recurrent) if it is:

- irreducible;
- recurrent; and,
- has an invariant probability distribution.

Thus, we have the following result [**82**].

THEOREM 6.1.3 (Sufficient Conditions for Ergodicity). *Let* $X(t)$ *be the Markov process with tridiagonal generator* Q *given by* (6.1.1), *Markov semigroup* $\{P_{ij}(t)\}_{t \geq 0}$ *defined as:*

$$P_{ij}(t) = \mathbb{P}(X(t) = x_j \mid X(0) = x_i).$$

and invariant density ν_d *whose formula is given in* (6.1.4). *This process is ergodic if it satisfies the following conditions:*

(C1): *For every integer* i

$$Q_{i,i+1} > 0 \quad \text{and} \quad Q_{i,i-1} > 0$$

(C2): *The following series converges*

$$Z = \sum_{i \in \mathbb{Z}} (\nu_d)_i < \infty$$

Let $\pi_d = \nu_d/Z$. *Then we have that:*

$$\sum_{j\in\mathbb{Z}} |P_{ij}(t) - (\pi_d)_j| \to 0 \qquad \text{as } t\to\infty \text{ for any } i\in\mathbb{Z}.$$

Here is a simple sufficient condition for (C2) to hold.

LEMMA 6.1.4. *Assume that* ν_d *satisfies* (6.1.2) *with seed values* $(\nu_d)_0 = 1$ *and* $(\nu_d)_1 = Q_{0,1}/Q_{1,0}$. *If the following inequalities are satisfied:*

$$\lim_{k\to\infty} \frac{Q_{k,k+1}}{Q_{k+1,k}} < 1 \quad \text{and} \quad \lim_{k\to-\infty} \frac{Q_{k,k-1}}{Q_{k-1,k}} < 1$$

then ν_d *is normalizable.*

PROOF. This lemma is an application of the ratio test, which gives a sufficient condition to establish convergence of a series. From (6.1.4) the quotient this test uses can be written as a ratio of jump rates:

$$\frac{(\nu_d)_{k+1}}{(\nu_d)_k} = \frac{Q_{k,k+1}}{Q_{k+1,k}}.$$

Hence, the ratio test for convergence of the series $\sum_{i\in\mathbb{N}}(\nu_d)_i$ reduces to confirming that:

$$\lim_{k\to\infty} \frac{Q_{k,k+1}}{Q_{k+1,k}} < 1\,,$$

and similarly for $i < 0$. □

6.2. Stationary Density Accuracy

In this section, we use the formula in (6.1.4) to obtain explicit estimates for the error of the infinitesimal generator $\tilde{Q}_c$ (given in (2.3.10)) in approximating the true stationary density. In particular, we show that the stationary density of $\tilde{Q}_c$ is second-order accurate.

PROPOSITION 6.2.1. *Consider the SDE* (2.3.3) *with* $\Omega = \mathbb{R}$. *Assume that:*

- $U(x)$ *is three times differentiable for all* $x \in \mathbb{R}$; *and*

$$\operatorname{sign}(x)U'(x) \to \infty \quad \text{as } |x| \to \infty.$$

- $M(x)$ *is differentiable and* $M(x) > 0$, *for all* $x \in \mathbb{R}$.

Let $\nu(x) = \exp(-U(x))$. *Suppose that the spatial step size is constant and that* $x_i = ih$ *for all* $i \in \mathbb{Z}$. *Consider the second-order scheme* (2.3.10) *with off-diagonal entries given by:*

$$(\tilde{Q}_c)_{i,i+1} = \frac{1}{h^2}\frac{M_i + M_{i+1}}{2}\exp\left(-U_i'\frac{h}{2}\right), \quad (\tilde{Q}_c)_{i,i-1} = \frac{1}{h^2}\frac{M_i + M_{i-1}}{2}\exp\left(U_i'\frac{h}{2}\right).$$

Let ν_c *be an invariant density of* Q_c *normalized so that* $\nu_c(x_0) = \nu(x_0)$. *Then,*

$$\sup_{x_i\in S} |\nu(x_i) - \nu_c(x_i)| < Ch^2 \tag{6.2.1}$$

where

$$C = \frac{1}{12}\sup_x \nu(x)|x| \sup_{\xi<|x|}|U'''(\xi)| \exp\left(\frac{h^2}{12}|x| \sup_{\xi<|x|}|U'''(\xi)|\right).$$

PROOF. First, we apply the test given in Lemma 6.1.4 to check that an invariant density of Q_c is normalizable. The hypothesis on $U'(x)$ implies that:

$$\lim_{k\to\infty}\frac{(\tilde{Q}_c)_{k,k+1}}{(\tilde{Q}_c)_{k+1,k}}=\lim_{k\to\infty}\exp\left(-\frac{h}{2}(U'_k+U'_{k+1})\right)=0$$

and similarly for $k<0$. Thus, $\tilde{Q}_c$ has a stationary density, and by Theorem 6.1.3, defines an ergodic process.

Next we use (6.1.4) to quantify the pointwise error in the approximation to the density $\nu(x)$. Let $\tilde{\nu}_c(x)$ be a function proportional to the stationary density of $\tilde{Q}_c$ such that $(\tilde{\nu}_c)_0=1$. (This density will be rescaled in order to obtain $\nu_c(x)$.) Substituting the rates appearing in (2.3.10) into (6.1.4) and simplifying yields:

$$\tilde{\nu}_c(x_i)=\begin{cases}\exp\left(-\frac{h}{2}\sum_{k=0}^{|i|-1}(U'_k+U'_{k+1})\right) & \text{if } i>0\\ \exp\left(\frac{h}{2}\sum_{k=0}^{|i|-1}(U'_{-k}+U'_{-k-1})\right) & \text{if } i<0\\ 1 & \text{if } i=0\end{cases}$$

We consider two cases depending on the value of x_i. Set $\nu_c=\tilde{\nu}_c/\exp(U_0)$, so that $\nu_c(x_0)=\nu(x_0)$.

If $x_i>0$: We have that,

$$\nu_c(x_i)=\exp\left(-\int_0^{x_i}U'(s)ds-U_0+\epsilon_h^+\right)=\exp\left(-U_i+\epsilon_h^+\right),$$

where we have introduced

$$\begin{aligned}\epsilon_h^+&=\int_0^{x_i}U'(s)ds-\frac{h}{2}\sum_{k=0}^{i-1}(U'_k+U'_{k+1})\\&=\sum_{k=0}^{i-1}\underbrace{\left(\int_{x_k}^{x_{k+1}}U'(s)ds-\frac{h}{2}(U'_k+U'_{k+1})\right)}_{\delta_k^+}=\sum_{k=0}^{i-1}\delta_k^+\end{aligned}$$

Use integration by parts twice to obtain the following expression for δ_k^+:

$$\delta_k^+=\frac{1}{2}\int_{x_k}^{x_{k+1}}\left((s-x_{k+\frac{1}{2}})^2-\left(\frac{h}{2}\right)^2\right)U'''(s)ds=-\frac{h^3}{12}U'''(\xi_k^+)$$

where $x_{k+\frac{1}{2}}=(x_k+x_{k+1})/2$ and $x_k\le\xi_k^+\le x_{k+1}$. Next we apply the elementary inequality:

$$|e^a-1|\le e^{|a|}-1\le|a|e^{|a|}\quad\forall\, a\in\mathbb{R}\tag{6.2.2}$$

to obtain

$$\begin{aligned}|\nu(x_i)-\nu_c(x_i)|&=\nu(x_i)\left|\exp\left(-\frac{h^3}{12}\sum_{k=0}^{i-1}U'''(\xi_k^+)\right)-1\right|\\&\le\frac{h^2}{12}\nu(x_i)|x_i|\sup_{\xi<|x_i|}|U'''(\xi)|\exp\left(\frac{h^2}{12}|x_i|\sup_{\xi<|x_i|}|U'''(\xi)|\right).\end{aligned}$$

If $x_i < 0$: We have that,

$$\nu_c(x_i) = \exp\left(\int_{x_i}^{0} U'(s)ds - U_0 + \epsilon_h^-\right) = \exp\left(-U_i + \epsilon_h^-\right) ,$$

and as in the previous case,

$$\epsilon_h^- = -\sum_{k=0}^{i-1} \frac{h^3}{12} U'''(\xi_k^-)$$

where $\xi_k^- \in [x_{k-1}, x_k]$. Applying (6.2.2) we obtain,

$$\begin{aligned} |\nu(x_i) - \nu_c(x_i)| &= \nu(x_i) \left| 1 - \exp\left(-\frac{h^3}{12}\sum_{k=0}^{i-1} U'''(\xi_k^-)\right)\right| \\ &\le \frac{h^2}{12}\nu(x_i)|x_i| \sup_{\xi<|x_i|} |U'''(\xi)| \exp\left(\frac{h^2}{12}|x_i| \sup_{\xi<|x_i|} |U'''(\xi)|\right) . \end{aligned}$$

Taking the sup over x_i in the last inequalities in each case gives the desired estimate. □

6.3. Exit Probability

Here we demonstrate how to compute the exit probability or committor function of the approximation induced by Q. As a side remark, the committor function can be used to analytically describe the statistical properties of trajectories between two metastable states of an SDE [**35**–**37**, **156**]. Given 'reactant' and 'product' states at the points a and b (resp.) with $a < b$, the committor function is defined as the probability that the process initiated at a point $x \in [a, b]$ reacts in the sense that it first hits b before it hits a. Assume that the reactant and product states are the left and right endpoints of the grid S and suppose that the grid points are arranged so that:

$$x_0 = a < \cdots < x_N = b$$

Let $q_d = \{(q_d)_i\}_{i=0}^N$ denote the committor function associated to the generator Q which satisfies:

$$\begin{cases} (Qq_d)_i = 0 & \text{if } 0 < i < N , \\ (q_d)_0 = 0 , \ (q_d)_N = 1 \end{cases} \tag{6.3.1}$$

The following proposition gives a formula for q_d in terms of the jump rates and invariant density ν_d of the approximation.

Proposition 6.3.1 (Formula for Exit Probability). *The committor function which solves* (6.3.1) *can be written as:*

$$(q_d)_i = \begin{cases} Z \displaystyle\sum_{j=0}^{i-1} \frac{(\nu_d)_1 Q_{1,0}}{(\nu_d)_{j+1} Q_{j+1,j}} & \textit{if } 1 \le i \le N \\ 0 & \textit{if } i = 0 \end{cases} \tag{6.3.2}$$

where Z is a constant selected so that $(q_d)_N = 1$.

PROOF. Let $\tilde{q}_d$ solve the recurrence relation in (6.3.1), which implies that

$$Q_{i,i+1}((\tilde{q}_d)_{i+1} - (\tilde{q}_d)_i) = Q_{i,i-1}((\tilde{q}_d)_i - (\tilde{q}_d)_{i-1}) \quad (\text{for } i \geq 1) \tag{6.3.3}$$

with seed values $(\tilde{q}_d)_0 = 0$ and $(\tilde{q}_d)_1 = 1$. Then $q_d = \tilde{q}_d/(\tilde{q}_d)_N$ solves the same recurrence relation with the desired boundary conditions: $(q_d)_0 = 0$ and $(q_d)_N = 1$. Hence, this proof finds a solution to (6.3.3). Multiply (6.3.3) by $(\nu_d)_i$ and use ν_d-symmetry to obtain:

$$(\nu_d)_{i+1}Q_{i+1,i}((\tilde{q}_d)_{i+1} - (\tilde{q}_d)_i) = (\nu_d)_i Q_{i,i-1}((\tilde{q}_d)_i - (\tilde{q}_d)_{i-1})$$

Sum these equations over the index i from 1 to any $k \geq 1$ and use summation by parts to obtain:

$$\sum_{i=1}^{k}(\nu_d)_{i+1}Q_{i+1,i}((\tilde{q}_d)_{i+1} - (\tilde{q}_d)_i) = \sum_{i=1}^{k}(\nu_d)_i Q_{i,i-1}((\tilde{q}_d)_i - (\tilde{q}_d)_{i-1})$$

$$\sum_{i=2}^{k+1}(\nu_d)_i Q_{i,i-1}((\tilde{q}_d)_i - (\tilde{q}_d)_{i-1}) = \sum_{i=1}^{k}(\nu_d)_i Q_{i,i-1}((\tilde{q}_d)_i - (\tilde{q}_d)_{i-1})$$

$$(\nu_d)_{k+1}Q_{k+1,k}((\tilde{q}_d)_{k+1} - (\tilde{q}_d)_k) = (\nu_d)_1 Q_{1,0}((\tilde{q}_d)_1 - (\tilde{q}_d)_0).$$

Combine this result with a telescoping sum to obtain:

$$\begin{aligned}
(\tilde{q}_d)_{k+1} &= (\tilde{q}_d)_{k+1} - (\tilde{q}_d)_0 + (\tilde{q}_d)_0 \\
&= \sum_{j=0}^{k}((\tilde{q}_d)_{j+1} - (\tilde{q}_d)_j) + (\tilde{q}_d)_0 \\
&= \sum_{j=0}^{k}\frac{(\nu_d)_1 Q_{1,0}}{(\nu_d)_{j+1}Q_{j+1,j}}((\tilde{q}_d)_1 - (\tilde{q}_d)_0) + (\tilde{q}_d)_0 \quad \text{for } k \geq 0.
\end{aligned}$$

Substituting the seed values $(\tilde{q}_d)_0 = 0$ and $(\tilde{q}_d)_1 = 1$ gives the solution to (6.3.3). The desired formula is obtained by setting $q_d = \tilde{q}_d/(\tilde{q}_d)_N$ or $Z = 1/(\tilde{q}_d)_N$ in (6.3.2). □

6.4. Mean First Passage Time

A similar recurrence relation can be derived for the mean first passage time (MFPT) of the approximation from any point in $[a,b]$ to a or b. To derive this recurrence, we reuse the grid constructed in §6.3. Define τ as

$$\tau = \min\{t \geq 0 \mid X(t) \in \{a,b\}\ ,\ X(0) = x_i\}.$$

Let $u_d = \{(u_d)_i\}_{i=0}^N$ be the MFPT, i.e. $u_d(x_i) = \mathbb{E}_{x_i}(\tau)$, which satisfies

$$\begin{cases} (Qu_d)_i = -1 & \text{if } 0 < i < N\ , \\ (u_d)_0 = (u_d)_N = 0 \end{cases} \tag{6.4.1}$$

The following proposition gives a formula for u_d in terms of the jump rates and invariant density ν_d of the approximation.

PROPOSITION 6.4.1 (Formula for MFPT). *The MFPT which solves (6.4.1) can be written as:*

$$(6.4.2)\qquad (u_d)_i = \begin{cases} Z\sum_{j=0}^{i-1} \frac{(\nu_d)_1 Q_{1,0}}{(\nu_d)_{j+1}Q_{j+1,j}} - \sum_{j=0}^{k}\sum_{i=1}^{j} \frac{(\nu_d)_i}{(\nu_d)_{j+1}Q_{j+1,j}} & \text{if } 0 < i < N \\ 0 & \text{if } i = 0 \text{ or } N \end{cases}$$

where Z is a constant selected so that $(u_d)_N = 0$.

PROOF. This proof is similar to the proof of Proposition 6.3.1, except that we do not introduce an auxiliary solution. Multiply (6.4.1) by $(\nu_d)_i$ and use ν_d-symmetry to obtain:

$$(\nu_d)_{i+1}Q_{i+1,i}((u_d)_{i+1} - (u_d)_i) - (\nu_d)_i Q_{i,i-1}((u_d)_i - (u_d)_{i-1}) = -(\nu_d)_i$$

Sum these equations over the index i from 1 to any $k \geq 1$ and use summation by parts to obtain:

$$\sum_{i=1}^{k}(\nu_d)_{i+1}Q_{i+1,i}((u_d)_{i+1} - (u_d)_i) = \sum_{i=1}^{k}(\nu_d)_i Q_{i,i-1}((u_d)_i - (u_d)_{i-1}) - \sum_{i=1}^{k}(\nu_d)_i$$

$$\sum_{i=2}^{k+1}(\nu_d)_i Q_{i,i-1}((u_d)_i - (u_d)_{i-1}) = \sum_{i=1}^{k}(\nu_d)_i Q_{i,i-1}((u_d)_i - (u_d)_{i-1}) - \sum_{i=1}^{k}(\nu_d)_i$$

$$(\nu_d)_{k+1}Q_{k+1,k}((u_d)_{k+1} - (u_d)_k) = (\nu_d)_1 Q_{1,0}((u_d)_1 - (u_d)_0) - \sum_{i=1}^{k}(\nu_d)_i.$$

Combine this result with a telescoping sum to obtain:

$$\begin{aligned}(u_d)_{k+1} &= (u_d)_{k+1} - (u_d)_0 + (u_d)_0 \\ &= \sum_{j=0}^{k}((u_d)_{j+1} - (u_d)_j) + (u_d)_0 \\ &= (u_d)_1 \sum_{j=0}^{k} \frac{(\nu_d)_1 Q_{1,0}}{(\nu_d)_{j+1}Q_{j+1,j}} - \sum_{j=0}^{k}\sum_{i=1}^{j} \frac{(\nu_d)_i}{(\nu_d)_{j+1}Q_{j+1,j}}\end{aligned}$$

To enforce the boundary condition at $x_N = b$ we solve

$$(u_d)_N = (u_d)_1 \sum_{j=0}^{N-1} \frac{(\nu_d)_1 Q_{1,0}}{(\nu_d)_{j+1}Q_{j+1,j}} - \sum_{j=0}^{N-1}\sum_{i=1}^{j} \frac{(\nu_d)_i}{(\nu_d)_{j+1}Q_{j+1,j}} = 0$$

for $(u_d)_1$ to obtain

$$(u_d)_1 = \frac{\sum_{j=0}^{N-1}\sum_{i=1}^{j} \frac{(\nu_d)_i}{(\nu_d)_{j+1}Q_{j+1,j}}}{\sum_{j=0}^{N-1} \frac{(\nu_d)_1 Q_{1,0}}{(\nu_d)_{j+1}Q_{j+1,j}}}$$

which gives the desired formula with $Z = (u_d)_1$. □

By construction, the mean holding time of the generator Q_e in (2.3.16) at every grid point x_i is equal to the MFPT of $Y(t)$ to $(x_{i-1}, x_{i+1})^c$ conditional on $Y(0) = x_i$. Here we show that this property implies that the Markov process associated to Q_e also preserves the MFPT of $Y(t)$ to $(x_k, x_j)^c$ for any $x_k, x_j \in S$ with $x_k < Y(0) = x_i < x_j$. Abstractly speaking, this result manifests the additive property of MFPTs.

LEMMA 6.4.2. *Let Q_e denote the infinitesimal generator in* (2.3.16) *that is constructed using the exact jump rates and transition probabilities of Y. Let $v(x)$ be the exact solution to:*

$$Lv(x) = -1 \quad \text{and} \quad v(x_k) = v(x_j) = 0 \tag{6.4.3}$$

where L is the infinitesimal generator of the SDE given in (2.3.1), *and $x_k, x_j \in S$ with $x_k < x_j$. Then $v(x)$ satisfies:*

$$Q_e v(x_i) = -1 \, ,$$

for all grid points $x_i \in (x_k, x_j)$.

PROOF. The unique solution to the boundary value problem (6.4.3) is:

$$v(x) = -\int_{x_k}^{x} \int_{x_k}^{y} \frac{\nu(r)}{M(y)\nu(y)} dr dy + K_1 \int_{x_k}^{x} \frac{dy}{M(y)\nu(y)} \tag{6.4.4}$$

where K_1 is the following constant:

$$K_1 = \frac{\int_{x_k}^{x_j} \int_{x_k}^{y} \frac{\nu(r)}{M(y)\nu(y)} dr dy}{\int_{x_k}^{x_j} \frac{dy}{M(y)\nu(y)}} .$$

Using (2.3.16), a simple calculation shows that:

$$\begin{aligned}
(Q_e v)_i &= \frac{1}{u_i} \left(\frac{q_i - q_{i-1}}{q_{i+1} - q_{i-1}} v_{i+1} - v_i + \frac{q_{i+1} - q_i}{q_{i+1} - q_{i-1}} v_{i-1} \right) \\
&= \frac{1}{u_i} \frac{1}{q_{i+1} - q_{i-1}} \left[(q_{i+1} - q_{i-1})(v_{i-1} - v_i) + (q_i - q_{i-1})(v_{i+1} - v_{i-1}) \right] .
\end{aligned}$$

To simplify this expression, recall that the exact MFPT to $(x_{i-1}, x_{i+1})^c$ is given by:

$$u(x) = -\int_{x_{i-1}}^{x} \int_{x_{i-1}}^{y} \frac{\nu(r)}{M(y)\nu(y)} dr dy + \tilde{K}_1 \int_{x_{i-1}}^{x} \frac{dy}{M(y)\nu(y)} \tag{6.4.5}$$

where $\tilde{K}_1$ is the following constant:

$$\tilde{K}_1 = \frac{\displaystyle\int_{x_{i-1}}^{x_{i+1}} \int_{x_{i-1}}^{y} \frac{\nu(r)}{M(y)\nu(y)} dr dy}{\displaystyle\int_{x_{i-1}}^{x_{i+1}} \frac{dy}{M(y)\nu(y)}} .$$

To be sure, $u(x)$ equals $v(x)$ when $x_k = x_{i-1}$ and $x_j = x_{i+1}$. From (6.4.4) and (6.4.5) it follows that:

$$
\begin{aligned}
(Q_e v)_i &= \frac{1}{u_i} \frac{1}{q_{i+1} - q_{i-1}} \Bigg[(q_{i+1} - q_{i-1}) \int_{x_{i-1}}^{x_i} \int_{x_k}^{y} \frac{\nu(r)}{M(y)\nu(y)} dr dy \\
&\qquad\qquad - (q_i - q_{i-1}) \int_{x_{i-1}}^{x_{i+1}} \int_{x_k}^{y} \frac{\nu(r)}{M(y)\nu(y)} dr dy \Bigg] \\
&= \frac{1}{u_i} \Bigg[-u_i + u_i + \int_{x_{i-1}}^{x_i} \int_{x_k}^{y} \frac{\nu(r)}{M(y)\nu(y)} dr dy \\
&\qquad - \frac{q_i - q_{i-1}}{q_{i+1} - q_{i-1}} \int_{x_{i-1}}^{x_{i+1}} \int_{x_k}^{y} \frac{\nu(r)}{M(y)\nu(y)} dr dy \Bigg] \\
&= \frac{1}{u_i} \Bigg[-u_i + u_i + \int_{x_{i-1}}^{x_i} \int_{x_{i-1}}^{y} \frac{\nu(r)}{M(y)\nu(y)} dr dy \\
&\qquad - \left(\frac{\int_{x_{i-1}}^{x_{i+1}} \int_{x_{i-1}}^{y} \frac{\nu(r)}{M(y)\nu(y)} dr dy}{\int_{x_{i-1}}^{x_{i+1}} \frac{dy}{M(y)\nu(y)}} \right) \int_{x_{i-1}}^{x_i} \frac{dy}{M(y)\nu(y)} \Bigg] \\
&= -1
\end{aligned}
$$

as required. In other words, the error of Q_e in approximating $v(x_i)$ is zero. □

CHAPTER 7

Conclusion and Outlook

This paper presented numerical methods to simulate SDEs that allow the displacement of the approximation to be bounded uniformly and adaptively in space. In other words, the current state of the approximation is at most one spatial step size away from its previous state. The existence of such a uniform bound often stabilizes approximations for both finite and long-time simulations. To construct approximations with this feature, we considered spatial discretizations of the infinitesimal generator of the SDE, which constitutes a shift in perspective away from the traditional approach based on time discretization of realizations of the SDE. In order to obtain scalable approximations, we required that these spatial discretizations satisfy a realizability condition, which we introduced in §2.1. If this realizability condition holds, then the discretization is called a realizable discretization, and implies that the approximate solution defined by the spatial discretization can be represented in terms of a continuous-time random walk.

This walk can be simulated without time discretization error using the SSA method described in Algorithm 2.1. This algorithm inputs jump rates and jump states that come directly from a realizable discretization of the infinitesimal generator of the SDE. Since we use a Monte-Carlo method to compute an approximate solution, this approach applies to problems that practically speaking are beyond the reach of standard numerical PDE methods. In particular, this approach does not require discretizing any part of the domain of the SDE problem. Thus, a realizable discretization is much cheaper to implement in comparison with standard numerical PDE methods, which have a computational cost that is typically exponential with respect to the dimension of the problem.

However, constructing realizable discretizations on gridded state spaces turned out to be difficult. Our most general result on gridded state spaces is given in Prop. 2.8.2, which states that a locally consistent and realizable discretization is possible to construct on a gridded state space if the covariance matrix is weakly diagonally dominant (see Definition 2.8.1). This spatial discretization required using jumps to nearest and next to nearest neighbors. In order to relax this weak diagonal dominance requirement without introducing jumps to next to ... next to nearest neighbors, we introduced the notion of a gridless state space (see Definition 2.2.1). With this state space, we were able to construct realizable discretizations for quite general SDEs. These discretizations were based on a rank-one expansion of the infinitesimal covariance matrix of the SDE. Properties of our most general realizable discretization are described in Prop. 2.7.1.

The numerical tests in Chapter 3 illustrated properties of these approximations, which distinguish them from existing methods. In several test problems, we found that the mean holding time of our approximation in any state automatically adapts to the stiffness of the SDE coefficients. For example, when the noise coefficient matrix was constant, we found that the mean holding time of the approximation at any state was inversely proportional to the magnitude of the drift at that state. As we illustrated in the Cox-Ingersoll-Ross and Lotka-Volterra SDE problems, the spatial step size itself can be adapted to the stiffness of the SDE coefficients. These examples showed that adaptive mesh refinement led to more accurate approximations in SDE problems with multiplicative noise, and they also showed (as expected) that this approximation never produces moves outside the domain of definition of the SDE. This property is a consequence of the jumps in the infinitesimal generator of the approximation being defined to lie in the domain of the SDE. Numerical tests on planar diffusions illustrated that our approximation is stable and accurate for SDEs with locally Lipschitz continuous drift fields, non-symmetric infinitesimal generators, and coefficients with internal discontinuities or singularities. We also considered a Brownian dynamics simulation of a cluster of 13 particles immersed in an unbounded solvent. The particle dynamics was given by gradient dynamics plus multiplicative noise due to the presence of the solvent. This example underscored that the spatial step size of our approximation can be adapted to a characteristic length scale in the SDE problem – namely, the Lennard-Jones radius appearing in the potential energy of the system. In this example, resolving this length scale is a necessary condition for accuracy.

In Chapter 4, we analyzed the properties of a second-order accurate, realizable discretization on a gridded state space with fixed spatial step size. The main issues in this analysis were that the state space (and hence the grid) may not be finite-dimensional, and that the linear operator associated to the approximation may not be bounded with respect to bounded grid functions. To deal with the unbounded nature of the state space and the infinitesimal generator, we used probabilistic tools to show that the approximation is stable. In particular, we showed that under a weak dissipativity and maximal/minimal polynomial growth conditions on the drift field, a sharp stochastic Lyapunov function was inherited by the approximation. A key result of this analysis is that the infinitesimal drift condition we derived is uniform in the spatial step size. We then used Harris theorem to prove that the approximation is geometrically ergodic with respect to a unique invariant probability measure.

This stochastic Lyapunov function was applied to solve a martingale problem associated to the approximation. Our analysis of a global semimartingale representation of sample paths quantified strong properties of the approximation, including the cost of the approximation as a function of the spatial step size parameter. To estimate the global error of the approximation, we obtained a stochastic representation of the global error as an expected value (over the approximation) of a function, which depends on the derivatives of the solution being approximated [**34**]. Our assumptions on the SDE implied that this solution had four bounded derivatives, and that the approximation had a stochastic Lyapunov function. With these two ingredients, we were able to expand this stochastic representation and obtain a global error estimate. This estimate revealed that the accuracy of our approximation at finite (and infinite-time) is determined by the local accuracy of the infinitesimal generator associated to the approximation.

A realizable discretization based on variable step sizes was analyzed in Chapter 5. Specifically, we considered a realizable discretization that uses randomly rescaled spatial increments, which mimic the effect of a state-dependent diffusion matrix. The main issue in this analysis turned out to be lack of irreducibility in the standard sense, and a related lack of a strong Feller property. These issues were not surprising since the approximation is a pure jump process with a countable state space, even though it may not be confined to a grid. Therefore the transition probabilities of the approximation were not absolutely continuous with respect to Lebesgue measure, and the underlying noise in the approximation did not have a regularizing effect. That said, we were still able to obtain a sharp stochastic Lyapunov function for the approximation, prove finite-time weak accuracy, prove existence of an invariant probability measure, and prove accuracy with respect to equilibrium expectations. The latter results required that the semigroup of the approximation to be Feller. To obtain this property, we mollified the infinitesimal generator of the approximation such that its jump rates were bounded but, the approximation still had a sharp stochastic Lyapunov function.

In Chapter 6 we analyzed the one-dimensional case where the infinitesimal generators of our approximations admit an infinite, tridiagonal matrix representation. By solving tridiagonal systems of equations, we verified some of the n-dimensional theory on gridded state spaces, and obtained explicit expressions for the stationary density, mean first passage time, and exit probability for various, one-dimensional approximations. These formulas were also used in the numerical tests in Chapter 3 to validate our approximations.

Let us finish by briefly describing some directions for future work. It would be interesting to construct the jumps appearing in (2.7.2) to reduce the computational cost of the approximation for high-dimensional SDE problems. As we discussed in §2.6, even though the computational cost of a realizable discretization is much less than exponential with respect to system size, the number of jumps used in our infinitesimal generators is currently dependent on dimension. In particular, recall that we used jumps based on a rank-one expansion of the infinitesimal covariance matrix given in (2.7.1). If this infinitesimal covariance matrix has full rank, then the number of terms in this rank-one expansion depends on dimension, and hence, the number of jumps used by the approximation per step depends on dimension. To overcome this dependence, and further reduce the computational cost of the approximation, we aim to replace (2.7.1) by the best rank-k approximation of $M(x)$, where $k \ll n$. Randomized numerical linear algebra methods for computing the top k eigenvalues/eigenvectors of a matrix may be particularly useful for estimating this best rank-k approximation [**57**].

Accuracy of realizable discretizations is currently limited to second-order. In the finite-dimensional, gridded state space context, this issue is related to the lack of higher-order, monotone approximations for elliptic differential operators. While higher-order accurate, realizable discretizations might be impractical to construct on gridded state spaces, it might be possible to construct such approximations on gridless state spaces. As an alternative to higher-order realizable discretizations, adaptive mesh refinement might offer better value in terms of accuracy. We stress that the main point of adaptivity is to increase accuracy, since stability of our approximations typically comes from a bound on the jump size of the approximation. Recall that the infinitesimal generator of the approximation given in (2.7.2) inputs forward/backward fields of spatial step sizes in each component. It would be interesting to specify these fields so that this variable step size approximation is more accurate than a fixed step size approximation, though our testing suggests that these techniques/results might be problem-specific.

In addition to reducing spatial discretization error, it would also be useful to develop variance-reduction methods adapted to our approximations, e.g., Multi-Level Monte-Carlo [**6**]. Another direction of development is to relax our (strong) requirement that the infinitesimal generator associated to the SDE is of elliptic type. Relaxing this requirement though introduces a spectrum of SDE problems: from those that satisfy Hörmander's condition and which have a smooth density with respect to Lebesgue measure to purely deterministic ODE problems. Accuracy/stability in these situations may require constructing realizable discretizations that have minimal numerical diffusion as described in [**20**]. In addition to the above directions, future work will also consider more applications and extensions of these ideas to SDEs with more general interior/boundary conditions [**7, 11, 33, 43, 91, 149**]. As a step in this direction, in Refs. [**13**] and [**16**] the techniques in this paper are extended to reflecting diffusions with oblique derivative boundary conditions and sticky boundary conditions, respectively.

Bibliography

[1] C. F. Abrams and E. Vanden-Eijnden, *Large-scale conformational sampling of proteins using temperature-accelerated molecular dynamics*, Proceedings of the National Academy of Sciences **107** (2010), no. 11, 4961–4966, `http://www.pnas.org/content/107/11/4961.abstract`.

[2] L. J. S. Allen, *An introduction to stochastic epidemic models*, Mathematical epidemiology, Lecture Notes in Math., vol. 1945, Springer, Berlin, 2008, pp. 81–130, DOI 10.1007/978-3-540-78911-6_3. MR2428373

[3] M. P. Allen and D. J. Tildesley, *Computer simulation of liquids*, Clarendon Press, 1987.

[4] R. J. Allen, D. Frenkel, and P. R. Ten Wolde, *Simulating rare events in equilibrium or nonequilibrium stochastic systems*, J Chem Phys **124** (2006), 024102, `http://jcp.aip.org/resource/1/jcpsa6/v124/i2/p024102_s1`.

[5] L. Andersen, P. Jäckel, and C. Kahl, *Simulation of square-root processes*, Encyclopedia of Quantitative Finance (2010).

[6] D. F. Anderson and D. J. Higham, *Multilevel Monte Carlo for continuous time Markov chains, with applications in biochemical kinetics*, Multiscale Model. Simul. **10** (2012), no. 1, 146–179, DOI 10.1137/110840546. MR2902602

[7] R. Atar and A. Budhiraja, *On the multi-dimensional skew Brownian motion*, Stochastic Process. Appl. **125** (2015), no. 5, 1911–1925, DOI 10.1016/j.spa.2014.12.001. MR3315617

[8] H. C. Berg, *Random walks in biology*, Princeton University Press, Princeton, NJ, 1983. MR835706

[9] M. Bertoldi and L. Lorenzi, *Estimates of the derivatives for parabolic operators with unbounded coefficients*, Trans. Amer. Math. Soc. **357** (2005), no. 7, 2627–2664 (electronic), DOI 10.1090/S0002-9947-05-03781-5. MR2139521

[10] E. G. Boman, D. Chen, O. Parekh, and S. Toledo, *On factor width and symmetric H-matrices*, Linear Algebra Appl. **405** (2005), 239–248, DOI 10.1016/j.laa.2005.03.029. MR2148173

[11] M. Bossy, E. Gobet, and D. Talay, *A symmetrized Euler scheme for an efficient approximation of reflected diffusions*, J. Appl. Probab. **41** (2004), no. 3, 877–889. MR2074829

[12] N. Bou-Rabee, *Time integrators for molecular dynamics*, Entropy **16** (2014), 138–162, `http://www.crab.rutgers.edu/~nb361/mypapers/Bo2014.pdf`.

[13] N. Bou-Rabee, *Markov Chain Approximation method for reflected diffusions.*

[14] N. Bou-Rabee, A. Donev, and E. Vanden-Eijnden, *Metropolis integration schemes for self-adjoint diffusions*, Multiscale Model. Simul. **12** (2014), no. 2, 781–831, DOI 10.1137/130937470. MR3223927

[15] N. Bou-Rabee and M. Hairer, *Nonasymptotic mixing of the MALA algorithm*, IMA J. Numer. Anal. **33** (2013), no. 1, 80–110, DOI 10.1093/imanum/drs003. MR3020951

[16] N. Bou-Rabee and M. Holmes-Cerfon, *Sticky diffusions and their numerical solution.* In preparation.

[17] N. Bou-Rabee and H. Owhadi, *Long-run accuracy of variational integrators in the stochastic context*, SIAM J. Numer. Anal. **48** (2010), no. 1, 278–297, DOI 10.1137/090758842. MR2608370

[18] N. Bou-Rabee and E. Vanden-Eijnden, *Pathwise accuracy and ergodicity of metropolized integrators for SDEs*, Comm. Pure Appl. Math. **63** (2010), no. 5, 655–696, DOI 10.1002/cpa.20306. MR2583309

[19] N. Bou-Rabee and E. Vanden-Eijnden, *A patch that imparts unconditional stability to explicit integrators for Langevin-like equations*, J. Comput. Phys. **231** (2012), no. 6, 2565–2580, DOI 10.1016/j.jcp.2011.12.007. MR2881033

[20] M. Boué and P. Dupuis, *Markov chain approximations for deterministic control problems with affine dynamics and quadratic cost in the control*, SIAM J. Numer. Anal. **36** (1999), no. 3, 667–695, DOI 10.1137/S0036142997323521. MR1681057

[21] T. Britton, *Stochastic epidemic models: A survey*, Math. Biosci. **225** (2010), no. 1, 24–35, DOI 10.1016/j.mbs.2010.01.006. MR2642269

[22] A. Brünger, C. L. Brooks, and M. Karplus, *Stochastic boundary conditions for molecular dynamics simulations of ST2 water*, Chem. Phys. Lett. **105** (1984), 495–500.

[23] S. Cerrai, *Second order PDE's in finite and infinite dimension*, Lecture Notes in Mathematics, vol. 1762, Springer-Verlag, Berlin, 2001. A probabilistic approach. MR1840644

[24] M.-F. Chen, *Equivalence of exponential ergodicity and L^2-exponential convergence for Markov chains*, Stochastic Process. Appl. **87** (2000), no. 2, 281–297, DOI 10.1016/S0304-4149(99)00114-3. MR1757116

[25] M. Chopra and R. G. Larson, *Brownian dynamics simulations of isolated polymer molecules in shear flow near adsorbing and nonadsorbing surfaces*, J Rheol **46** (2002), 831, `http://journalofrheology.org/resource/1/jorhd2/v46/i4/p831_s1`.

[26] P. G. Ciarlet, *Discrete maximum principle for finite-difference operators*, Aequationes Math. **4** (1970), 338–352. MR0292317

[27] R. R. Coifman and S. Lafon, *Diffusion maps*, Appl. Comput. Harmon. Anal. **21** (2006), no. 1, 5–30, DOI 10.1016/j.acha.2006.04.006. MR2238665

[28] R. R. Coifman, S. Lafon, A. B. Lee, M. Maggioni, B. Nadler, F. Warner, and S. W. Zucker, *Geometric diffusions as a tool for harmonic analysis and structure definition of data: Diffusion maps*, Proceedings of the National Academy of Sciences of the United States of America **102** (2005), no. 21, 7426–7431.

[29] V. Colizza, A. Barrat, M. Barthélemy, and A. Vespignani, *The role of the airline transportation network in the prediction and predictability of global epidemics*, Proceedings of the National Academy of Sciences of the United States of America **103** (2006), no. 7, 2015–2020.

[30] J. C. Cox, J. E. Ingersoll Jr, and S. A. Ross, *A theory of the term structure of interest rates*, Econometrica: Journal of the Econometric Society (1985), 385–407.

[31] J. J. de Pablo, *Coarse-grained simulations of macromolecules: From DNA to nanocomposites*, Annu Rev Phys Chem **62** (2011), 555–574, `http://www.annualreviews.org/doi/abs/10.1146/annurev.physchem.040808.090412`.

[32] S. Delong, F. B. Usabiaga, R. Delgado-Busalioni, B. E. Griffith, and A. Donev, *Brownian dynamics without Green's functions*, J Chem Phys **140** (2014), no. 13, 134110.

[33] P. Dupuis and H. Ishii, *SDEs with oblique reflection on nonsmooth domains*, Ann. Probab. **21** (1993), no. 1, 554–580. MR1207237

[34] P. Dupuis and M. R. James, *Rates of convergence for approximation schemes in optimal control*, SIAM J. Control Optim. **36** (1998), no. 2, 719–741, DOI 10.1137/S0363012994267789. MR1616550

[35] W. E and E. Vanden-Eijnden, *Metastability, conformation dynamics, and transition pathways in complex systems*, Multiscale modelling and simulation, Lect. Notes Comput. Sci. Eng., vol. 39, Springer, Berlin, 2004, pp. 35–68, DOI 10.1007/978-3-642-18756-8_3. MR2089952

[36] W. E and E. Vanden-Eijnden, *Towards a theory of transition paths*, J. Stat. Phys. **123** (2006), no. 3, 503–523, DOI 10.1007/s10955-005-9003-9. MR2252154

[37] Weinan E and Eric Vanden-Eijnden, *Transition-path theory and path-finding algorithms for the study of rare events*, Annual Review of Physical Chemistry **61** (2010), 391–420, `http://www.annualreviews.org/doi/abs/10.1146/annurev.physchem.040808.090412`.

[38] J.-P. Eckmann and M. Hairer, *Non-equilibrium statistical mechanics of strongly anharmonic chains of oscillators*, Comm. Math. Phys. **212** (2000), no. 1, 105–164, DOI 10.1007/s002200000216. MR1764365

[39] J.-P. Eckmann, C.-A. Pillet, and L. Rey-Bellet, *Non-equilibrium statistical mechanics of anharmonic chains coupled to two heat baths at different temperatures*, Comm. Math. Phys. **201** (1999), no. 3, 657–697, DOI 10.1007/s002200050572. MR1685893

[40] T. C. Elston and C. R. Doering, *Numerical and analytical studies of nonequilibrium fluctuation-induced transport processes*, J Stat Phys **83** (1996), no. 3-4, 359–383.

[41] S. N. Ethier and T. G. Kurtz, *Markov processes: characterization and convergence*, vol. 282, John Wiley & Sons, 2009.

[42] K. Feng and D. L. Wang, *Dynamical systems and geometric construction of algorithms*, Computational mathematics in China, Contemp. Math., vol. 163, Amer. Math. Soc., Providence, RI, 1994, pp. 1–32, DOI 10.1090/conm/163/01547. MR1276072

[43] E. R. Fernholz, T. Ichiba, and I. Karatzas, *Two Brownian particles with rank-based characteristics and skew-elastic collisions*, Stochastic Process. Appl. **123** (2013), no. 8, 2999–3026, DOI 10.1016/j.spa.2013.03.019. MR3062434

[44] D. Frenkel and B. Smit, *Understanding molecular simulation: From algorithms to applications, second edition*, Academic Press, 2002.

[45] A. Furukawa and H. Tanaka, *Key role of hydrodynamic interactions in colloidal gelation*, Phys Rev Lett **104** (2010), no. 24, 245702.

[46] J. G. Gaines and T. J. Lyons, *Variable step size control in the numerical solution of stochastic differential equations*, SIAM J. Appl. Math. **57** (1997), no. 5, 1455–1484, DOI 10.1137/S0036139995286515. MR1470933

[47] D. T. Gillespie, *Exact stochastic simulation of coupled chemical reactions*, Journal of Physical Chemistry **81** (1977), no. 25, 2340–2361.

[48] E. Gobet and S. Menozzi, *Exact approximation rate of killed hypoelliptic diffusions using the discrete Euler scheme*, Stochastic Process. Appl. **112** (2004), no. 2, 201–223, DOI 10.1016/j.spa.2004.03.002. MR2073411

[49] Emmanuel Gobet and Stéphane Menozzi, *Discrete sampling of functionals of Itô processes*, Séminaire de probabilités XL, Springer, 2007, pp. 355–374.

[50] E. Gobet and S. Menozzi, *Stopped diffusion processes: boundary corrections and overshoot*, Stochastic Process. Appl. **120** (2010), no. 2, 130–162, DOI 10.1016/j.spa.2009.09.014. MR2576884

[51] M. D. Graham, *Fluid dynamics of dissolved polymer molecules in confined geometries*, Annual review of fluid mechanics. Volume 43, 2011, Annu. Rev. Fluid Mech., vol. 43, Annual Reviews, Palo Alto, CA, 2011, pp. 273–298, DOI 10.1146/annurev-fluid-121108-145523. MR2768018

[52] B. Gustafsson, H.-O. Kreiss, and J. Oliger, *Time-dependent problems and difference methods*, 2nd ed., Pure and Applied Mathematics (Hoboken), John Wiley & Sons, Inc., Hoboken, NJ, 2013. MR3235981

[53] M. Hairer, *Ergodic theory for stochastic PDEs*, Unpublished lecture notes (2008).

[54] M. Hairer and J. C. Mattingly, *Ergodicity of the 2D Navier-Stokes equations with degenerate stochastic forcing*, Ann. of Math. (2) **164** (2006), no. 3, 993–1032, DOI 10.4007/annals.2006.164.993. MR2259251

[55] M. Hairer and J. C. Mattingly, *Slow energy dissipation in anharmonic oscillator chains*, Comm. Pure Appl. Math. **62** (2009), no. 8, 999–1032, DOI 10.1002/cpa.20280. MR2531551

[56] Martin Hairer and Jonathan C. Mattingly, Seminar on Stochastic Analysis, Random Fields and Applications VI, Springer, 2011, pp. 109–117.

[57] N. Halko, P. G. Martinsson, and J. A. Tropp, *Finding structure with randomness: probabilistic algorithms for constructing approximate matrix decompositions*, SIAM Rev. **53** (2011), no. 2, 217–288, DOI 10.1137/090771806. MR2806637

[58] U. G. Haussmann and É. Pardoux, *Time reversal of diffusions*, Ann. Probab. **14** (1986), no. 4, 1188–1205. MR866342

[59] J. P. Hernández-Ortiz, J. J. de Pablo, and M. D. Graham, *Fast computation of many-particle hydrodynamic and electrostatic interactions in a confined geometry*, Phys Rev Lett **98** (2007), 140602, `http://prl.aps.org/abstract/PRL/v98/i14/e140602`.

[60] J. P. Hernández-Ortiz, H. Ma, J. J. de Pablo, and M. D. Graham, *Cross-stream-line migration in confined flowing polymer solutions: theory and simulation*, Phys Fluids **18** (2006), 123101, `http://pof.aip.org/resource/1/phfle6/v18/i12/p123101_s1`.

[61] ———, *$N \log N$ method for hydrodynamic interactions of confined polymer systems: Brownian dynamics*, J Chem Phys **125** (2006), 164906, `http://jcp.aip.org/resource/1/jcpsa6/v125/i16/p164906_s1`.

[62] ———, *Concentration distributions during flow of confined flowing polymer solutions at finite concentration: slit and grooved channel*, Korea-Aust Rhelo J **20** (2008), 143, `http://infosys.korea.ac.kr/PDF/KARJ/KR20/KR20-3-0143.pdf`.

[63] D. J. Higham, *Stochastic ordinary differential equations in applied and computational mathematics*, IMA J. Appl. Math. **76** (2011), no. 3, 449–474, DOI 10.1093/imamat/hxr016. MR2806005

[64] D. J. Higham, X. Mao, and A. M. Stuart, *Strong convergence of Euler-type methods for nonlinear stochastic differential equations*, SIAM J. Numer. Anal. **40** (2002), no. 3, 1041–1063 (electronic), DOI 10.1137/S0036142901389530. MR1949404

[65] C.-C. Hsieh and T.-H. Lin, *Simulation of conformational preconditioning strategies for electrophoretic stretching of DNA in a microcontraction*, Biomicrofluidics **5** (2011), no. 4, 044106, `http://bmf.aip.org/resource/1/biomgb/v5/i4/p044106_s1`.

[66] C.-C. Hsieh, T.-H. Lin, and C.-D. Huang, *Simulation guided design of a microfluidic device for electrophoretic stretching of DNA*, Biomicrofluidics **6** (2012), 044105, `http://bmf.aip.org/resource/1/biomgb/v6/i4/p044105_s1`.

[67] I. Huopaniemi, K. Luo, T. Ala-Nissila, and S.-C. Ying, *Langevin dynamics simulations of polymer translocation through nanopores*, J Chem Phys **125** (2006), 124901, `http://jcp.aip.org/resource/1/jcpsa6/v125/i12/p124901_s1`.

[68] J. S. Hur, E. S. G. Shaqfeh, H. P. Babcock, and S. Chu, *Dynamics and configurational fluctuations of single DNA molecules in linear mixed flows*, Phys Rev E **66** (2002), 011915, `http://www.sciencedirect.com/science/article/pii/S0377025702001325`.

[69] J. S. Hur, E. S. G. Shaqfeh, and R. G. Larson, *Brownian dynamics simulations of single DNA molecules in shear flow*, J Rheol **44** (2000), 713, `http://journalofrheology.org/resource/1/jorhd2/v44/i4/p713_s1`.

[70] M. Hutzenthaler, A. Jentzen, and P. E. Kloeden, *Strong and weak divergence in finite time of Euler's method for stochastic differential equations with non-globally Lipschitz continuous coefficients*, Proc. R. Soc. Lond. Ser. A Math. Phys. Eng. Sci. **467** (2011), no. 2130, 1563–1576, DOI 10.1098/rspa.2010.0348. MR2795791

[71] M. Hutzenthaler, A. Jentzen, and P. E. Kloeden, *Strong convergence of an explicit numerical method for SDEs with nonglobally Lipschitz continuous coefficients*, Ann. Appl. Probab. **22** (2012), no. 4, 1611–1641, DOI 10.1214/11-AAP803. MR2985171

[72] N. Ikeda and S. Watanabe, *Stochastic differential equations and diffusion processes*, 2nd ed., North-Holland Mathematical Library, vol. 24, North-Holland Publishing Co., Amsterdam; Kodansha, Ltd., Tokyo, 1989. MR1011252

[73] R. M. Jendrejack, J. J. de Pablo, and M. D. Graham, *Stochastic simulations of DNA in flow: Dynamics and the effects of hydrodynamic interaction*, J Chem Phys **116** (2002), 7752, `http://jcp.aip.org/resource/1/jcpsa6/v116/i17/p7752_s1`.

[74] R. M. Jendrejack, E. T. Dimalanta, D. C. Schwartz, M. D. Graham, and J. J. de Pablo, *DNA dynamics in a microchannel*, Phys Rev Lett **91** (2003), 038102, `http://prl.aps.org/abstract/PRL/v91/i3/e038102`.

[75] ———, *Effect of confinement on DNA dynamics in microfluidic devices*, J Chem Phys **119** (2003), 2894, `http://jcp.aip.org/resource/1/jcpsa6/v119/i2/p1165_s1`.

[76] R. M. Jendrejack, D. C. Schwartz, J. J. de Pablo, and M. D. Graham, *Shear-induced migration in flowing polymer solutions: simulation of long-chain deoxyribose nucleic acid in micro channels*, J Chem Phys **120** (2004), 2513, `http://jcp.aip.org/resource/1/jcpsa6/v120/i5/p2513_s1`.

[77] I. Karatzas and S. E. Shreve, *Methods of mathematical finance*, Applications of Mathematics (New York), vol. 39, Springer-Verlag, New York, 1998. MR1640352

[78] R. Khasminskii, *Stochastic stability of differential equations*, 2nd ed., Stochastic Modelling and Applied Probability, vol. 66, Springer, Heidelberg, 2012. With contributions by G. N. Milstein and M. B. Nevelson. MR2894052

[79] A. Khinchine, *Asymptotische gesetze der wahrscheinlichkeitsrechnun*, Springer, 1933.

[80] J. M. Kim and P. S. Doyle, *A Brownian dynamics-finite element method for simulating DNA electrophoresis in nonhomogeneous electric fields*, J Chem Phys **125** (2006), 074906, `http://web.mit.edu/doylegroup/pubs/JChemPhys_JuMin_06.pdf`.

[81] ———, *Design and numerical simulation of a DNA electrophoretic stretching device*, Lab on a Chip **7** (2007), no. 2, 213–225, `http://web.mit.edu/doylegroup/pubs/LOC_JuMin_07.pdf`.

[82] J. R. Kirkwood, *Markov processes*, Advances in Applied Mathematics, CRC Press, Boca Raton, FL, 2015. MR3328907

[83] F. C. Klebaner, *Introduction to stochastic calculus with applications*, 3rd ed., Imperial College Press, London, 2012. MR2933773

[84] P. E. Kloeden and E. Platen, *Numerical solution of stochastic differential equations*, Applications of Mathematics (New York), vol. 23, Springer-Verlag, Berlin, 1992. MR1214374

[85] I. Kontoyiannis and S. P. Meyn, *Spectral theory and limit theorems for geometrically ergodic Markov processes*, Ann. Appl. Probab. **13** (2003), no. 1, 304–362, DOI 10.1214/aoap/1042765670. MR1952001

[86] I. Kontoyiannis and S. P. Meyn, *Geometric ergodicity and the spectral gap of non-reversible Markov chains*, Probab. Theory Related Fields **154** (2012), no. 1-2, 327–339, DOI 10.1007/s00440-011-0373-4. MR2981426

[87] T. G. Kurtz, *Approximation of population processes*, CBMS-NSF Regional Conference Series in Applied Mathematics, vol. 36, Society for Industrial and Applied Mathematics (SIAM), Philadelphia, Pa., 1981. MR610982

[88] H. J. Kushner and P. G. Dupuis, *Numerical methods for stochastic control problems in continuous time*, Applications of Mathematics (New York), vol. 24, Springer-Verlag, New York, 1992. MR1217486

[89] H. J. Kushner, *Probability limit theorems and the convergence of finite difference approximations of partial differential equations*, J. Math. Anal. Appl. **32** (1970), 77–103; erratum and addendum, ibid. 34 (1971), 233. MR0292169

[90] H. J. Kushner, *Finite difference methods for the weak solutions of the Kolmogorov equations for the density of both diffusion and conditional diffusion processes*, J. Math. Anal. Appl. **53** (1976), no. 2, 251–265. MR0461683

[91] H. J. Kushner, *Probabilistic methods for finite difference approximations to degenerate elliptic and parabolic equations with Neumann and Dirichlet boundary conditions*, J. Math. Anal. Appl. **53** (1976), no. 3, 644–668. MR0400424

[92] H. J. Kushner, *Probability methods for approximations in stochastic control and for elliptic equations*, Academic Press, 1977.

[93] H. J. Kushner, *Numerical approximations for stochastic differential games*, SIAM J. Control Optim. **41** (2002), no. 2, 457–486 (electronic), DOI 10.1137/S0363012901389457. MR1920267

[94] H. J. Kushner, *Numerical approximations for stochastic differential games: the ergodic case*, SIAM J. Control Optim. **42** (2004), no. 6, 1911–1933 (electronic), DOI 10.1137/S0036301290140034. MR2080921

[95] H. J. Kushner, *Numerical approximations for stochastic systems with delays in the state and control*, Stochastics **78** (2006), no. 5, 343–376, DOI 10.1080/17442500600893480. MR2270942

[96] H. J. Kushner, *Numerical methods for controls for nonlinear stochastic systems with delays and jumps: applications to admission control*, Stochastics **83** (2011), no. 3, 277–310, DOI 10.1080/17442508.2011.552724. MR2810593

[97] H. J. Kushner and C. F. Yu, *Probability methods for the convergence of finite difference approximations to partial differential equations*, J. Math. Anal. Appl. **43** (1973), 603–625. MR0323125

[98] H. J. Kushner and C. F. Yu, *The approximate calculation of invariant measures of diffusions via finite difference approximations to degenerate elliptic partial differential equations*, J. Math. Anal. Appl. **51** (1975), no. 2, 359–367. MR0378121

[99] H. Lamba, J. C. Mattingly, and A. M. Stuart, *An adaptive Euler-Maruyama scheme for SDEs: convergence and stability*, IMA J. Numer. Anal. **27** (2007), no. 3, 479–506, DOI 10.1093/imanum/drl032. MR2337577

[100] R. G. Larson, *The rheology of dilute solutions of flexible polymers: Progress and problems*, J Rheol **49** (2005), 1–70, `http://journalofrheology.org/resource/1/jorhd2/v49/i1/p1_s1`.

[101] R. G. Larson, H. Hu, D. E. Smith, and S. Chu, *Brownian dynamics simulations of a DNA molecule in an extensional flow field*, J Rheol **43** (1999), 267, `http://journalofrheology.org/resource/1/jorhd2/v43/i2/p267_s1`.

[102] J. C. Latorre, P. Metzner, C. Hartmann, and C. Schütte, *A structure-preserving numerical discretization of reversible diffusions*, Commun. Math. Sci. **9** (2011), no. 4, 1051–1072, DOI 10.4310/CMS.2011.v9.n4.a6. MR2901816

[103] R. J. LeVeque, *Finite difference methods for ordinary and partial differential equations*, Society for Industrial and Applied Mathematics (SIAM), Philadelphia, PA, 2007. Steady-state and time-dependent problems. MR2378550

[104] K. K. Lin and L.-S. Young, *Nonequilibrium steady states for certain Hamiltonian models*, J. Stat. Phys. **139** (2010), no. 4, 630–657, DOI 10.1007/s10955-010-9958-z. MR2638931

[105] H. C. Loebl, R. Randel, S. P. Goodwin, and C. C. Matthai, *Simulation studies of polymer translocation through a channel*, Phys Rev E **67** (2003), no. 4, 041913, http://pre.aps.org/abstract/PRE/v67/i4/e041913.

[106] R. Lord, R. Koekkoek, and D. Van Dijk, *A comparison of biased simulation schemes for stochastic volatility models*, Quant. Finance **10** (2010), no. 2, 177–194, DOI 10.1080/14697680802392496. MR2642962

[107] L. Lorenzi and M. Bertoldi, *Analytical methods for Markov semigroups*, CRC Press, 2006.

[108] D. K. Lubensky and D. R. Nelson, *Driven polymer translocation through a narrow pore*, Biophysical Journal **77** (1999), no. 4, 1824–1838, http://www.sciencedirect.com/science/article/pii/S000634959977027X.

[109] R. S. Maier and D. L. Stein, *Transition-rate theory for nongradient drift fields*, Phys Rev Lett **69** (1992), no. 26, 3691, http://prl.aps.org/abstract/PRL/v69/i26/p3691_1.

[110] Robert S. Maier and Daniel L. Stein, *Escape problem for irreversible systems*, Phys Rev E **48** (1993), no. 2, 931, http://pre.aps.org/abstract/PRE/v48/i2/p931_1.

[111] R. S. Maier and D. L. Stein, *A scaling theory of bifurcations in the symmetric weak-noise escape problem*, J. Statist. Phys. **83** (1996), no. 3-4, 291–357, DOI 10.1007/BF02183736. MR1386348

[112] A. J. Majda and X. Wang, *Non-linear dynamics and statistical theories for basic geophysical flows*, Cambridge University Press, Cambridge, 2006. MR2241372

[113] R. Mannella, *Absorbing boundaries and optimal stopping in a stochastic differential equation*, Phys. Lett. A **254** (1999), no. 5, 257–262, DOI 10.1016/S0375-9601(99)00117-6. MR1687576

[114] X. Mao, *Stochastic differential equations and applications*, Elsevier, 2007.

[115] X. Mao, G. Marion, and E. Renshaw, *Environmental Brownian noise suppresses explosions in population dynamics*, Stochastic Process. Appl. **97** (2002), no. 1, 95–110, DOI 10.1016/S0304-4149(01)00126-0. MR1870962

[116] L. Maragliano and E. Vanden-Eijnden, *A temperature accelerated method for sampling free energy and determining reaction pathways in rare events simulations*, Chem Phys Lett **426** (2006), http://www.cims.nyu.edu/~eve2/cpl_langevin_temp.pdf.

[117] J. C. Mattingly, A. M. Stuart, and M. V. Tretyakov, *Convergence of numerical time-averaging and stationary measures via Poisson equations*, SIAM J. Numer. Anal. **48** (2010), no. 2, 552–577, DOI 10.1137/090770527. MR2669996

[118] S. Matysiak, A. Montesi, M. Pasquali, A. B. Kolomeisky, and C. Clementi, *Dynamics of polymer translocation through nanopores: Theory meets experiment*, Phys Rev Lett **96** (2006), no. 11, 118103, http://prl.aps.org/abstract/PRL/v96/i11/e118103.

[119] G. Metafune, D. Pallara, and E. Priola, *Spectrum of Ornstein-Uhlenbeck operators in L^p spaces with respect to invariant measures*, J. Funct. Anal. **196** (2002), no. 1, 40–60, DOI 10.1006/jfan.2002.3978. MR1941990

[120] P. Metzer, *Transition path theory for Markov processes*, Doctoral thesis, Free University Berlin, 1999.

[121] P. Metzner, C. Schütte, and E. Vanden-Eijnden, *Transition path theory for Markov jump processes*, Multiscale Model. Simul. **7** (2008), no. 3, 1192–1219, DOI 10.1137/070699500. MR2480117

[122] S. P. Meyn and R. L. Tweedie, *Stability of Markovian processes. III. Foster-Lyapunov criteria for continuous-time processes*, Adv. in Appl. Probab. **25** (1993), no. 3, 518–548, DOI 10.2307/1427522. MR1234295

[123] S. Meyn and R. L. Tweedie, *Markov chains and stochastic stability*, 2nd ed., Cambridge University Press, Cambridge, 2009. With a prologue by Peter W. Glynn. MR2509253

[124] G. N. Milstein and M. V. Tretyakov, *Stochastic numerics for mathematical physics*, Scientific Computation, Springer-Verlag, Berlin, 2004. MR2069903

[125] G. N. Milstein and M. V. Tretyakov, *Numerical integration of stochastic differential equations with nonglobally Lipschitz coefficients*, SIAM J. Numer. Anal. **43** (2005), no. 3, 1139–1154 (electronic), DOI 10.1137/040612026. MR2177799

[126] B. Nadler, S. Lafon, R. R. Coifman, and I. G. Kevrekidis, *Difusion maps, spectral clustering and reaction coordinates of dynamical systems*, Appl. Comput. Harmon. Anal. **21** (2006), no. 1, 113–127, DOI 10.1016/j.acha.2005.07.004. MR2238669

[127] H. C. Öttinger, *Stochastic processes in polymeric fluids*, Springer-Verlag, Berlin, 1996. Tools and examples for developing simulation algorithms. MR1383323

[128] R. G. Pinsky, *Positive harmonic functions and diffusion*, Cambridge Studies in Advanced Mathematics, vol. 45, Cambridge University Press, Cambridge, 1995. MR1326606

[129] E. Platen and D. Heath, *A benchmark approach to quantitative finance*, Springer Finance, Springer-Verlag, Berlin, 2006. MR2267213

[130] G. Da Prato and J. Zabczyk, *Ergodicity for infinite-dimensional systems*, London Mathematical Society Lecture Note Series, vol. 229, Cambridge University Press, Cambridge, 1996. MR1417491

[131] P. E. Protter, *Stochastic integration and differential equations*, 2nd ed., Applications of Mathematics (New York), vol. 21, Springer-Verlag, Berlin, 2004. Stochastic Modelling and Applied Probability. MR2020294

[132] G. C. Randall, K. M. Schultz, and P. S. Doyle, *Methods to electrophoretically stretch DNA: microcontractions, gels, and hybrid gel-microcontraction devices*, Lab on a Chip **6** (2006), no. 4, 516–525, `http://pubs.rsc.org/en/content/articlehtml/2006/lc/b515326c`.

[133] D. Revuz and M. Yor, *Continuous martingales and Brownian motion*, 3rd ed., Grundlehren der Mathematischen Wissenschaften [Fundamental Principles of Mathematical Sciences], vol. 293, Springer-Verlag, Berlin, 1999. MR1725357

[134] L. Rey-Bellet and L. E. Thomas, *Exponential convergence to non-equilibrium stationary states in classical statistical mechanics*, Comm. Math. Phys. **225** (2002), no. 2, 305–329, DOI 10.1007/s002200100583. MR1889227

[135] G. O. Roberts and J. S. Rosenthal, *Geometric ergodicity and hybrid Markov chains*, Electron. Comm. Probab. **2** (1997), no. 2, 13–25 (electronic), DOI 10.1214/ECP.v2-981. MR1448322

[136] G. O. Roberts and R. L. Tweedie, *Exponential convergence of Langevin distributions and their discrete approximations*, Bernoulli **2** (1996), no. 4, 341–363, DOI 10.2307/3318418. MR1440273

[137] G. O. Roberts and R. L. Tweedie, *Geometric convergence and central limit theorems for multidimensional Hastings and Metropolis algorithms*, Biometrika **83** (1996), no. 1, 95–110, DOI 10.1093/biomet/83.1.95. MR1399158

[138] J. Rotne and S. Prager, *Variational treatment of hydrodynamic interaction in polymers*, J Chem Phys **50** (1969), 4831, `http://jcp.aip.org/resource/1/jcpsa6/v50/i11/p4831_s1`.

[139] R. Rudnicki and K. Pichór, *Influence of stochastic perturbation on prey-predator systems*, Math. Biosci. **206** (2007), no. 1, 108–119, DOI 10.1016/j.mbs.2006.03.006. MR2311674

[140] J. M. Sanz-Serna and A. M. Stuart, *Ergodicity of dissipative differential equations subject to random impulses*, J. Differential Equations **155** (1999), no. 2, 262–284, DOI 10.1006/jdeq.1998.3594. MR1698555

[141] T. Schneider and E. Stoll, *Molecular-dynamics study of a three-dimensional one-component model for distortive phase transitions*, Physical Review B **17** (1978), 1302–1322.

[142] C. M. S. Schroeder, H. B. Babcock, E. S. G. Shaqfeh, and S. Chu, *Observation of polymer conformation hysteresis in extensional flow*, Science **301** (2003), 1515, `http://www.sciencemag.org/content/301/5639/1515.full.html`.

[143] C. M. S. Schroeder, E. S. G. Shaqfeh, and S. Chu, *The effect of hydrodynamic interactions on DNA dynamics in extensional flow: simulation and single molecule experiment*, Macromolecules **37** (2004), 9242, `http://www.scs.illinois.edu/schroeder/publications/effectofhydrodynamic.pdf`.

[144] C. Schütte, *Conformational dynamics: Modeling, theory, algorithm, and application to biomolecules*, Habilitation, Free University Berlin, 1999.

[145] Ch. Schütte, W. Huisinga, and P. Deuflhard, *Transfer operator approach to conformational dynamics in biomolecular systems*, Ergodic theory, analysis, and efficient simulation of dynamical systems, Springer, Berlin, 2001, pp. 191–223. MR1850307

[146] E. S. G. Shaqfeh, *The dynamics of single-molecule DNA in flow*, J Non-Newton Fluid **130** (2005), 1–28, `http://www.sciencedirect.com/science/article/pii/S0377025705001631`.

[147] Ya. G. Sinaĭ, *The limit behavior of a one-dimensional random walk in a random environment* (Russian, with English summary), Teor. Veroyatnost. i Primenen. **27** (1982), no. 2, 247–258. MR657919

[148] J. C. Strikwerda, *Finite difference schemes and partial differential equations*, 2nd ed., Society for Industrial and Applied Mathematics (SIAM), Philadelphia, PA, 2004. MR2124284

[149] D. W. Stroock and S. R. S. Varadhan, *Diffusion processes with boundary conditions*, Comm. Pure Appl. Math. **24** (1971), 147–225. MR0277037

[150] A. Szepessy, R. Tempone, and G. E. Zouraris, *Adaptive weak approximation of stochastic differential equations*, Comm. Pure Appl. Math. **54** (2001), no. 10, 1169–1214, DOI 10.1002/cpa.10000. MR1843985

[151] D. Talay, *Simulation and numerical analysis of stochastic differential systems : a review*, Probabilistic Methods in Applied Physics (P. Krèe and W. Wedig, eds.), vol. 451, Springer-Verlag, Berlin, 1995, pp. 54–96.

[152] ———, *Stochastic Hamiltonian systems: Exponential convergence to the invariant measure, and discretization by the implicit Euler scheme*, Markov Processes and Related Fields **8** (2002), 1–36, http://www-sop.inria.fr/members/Denis.Talay/fichiers-pdf/hamilton.pdf.

[153] D. Talay and L. Tubaro, *Expansion of the global error for numerical schemes solving stochastic differential equations*, Stochastic Anal. Appl. **8** (1990), no. 4, 483–509 (1991), DOI 10.1080/07362999008809220. MR1091544

[154] D. W. Trahan and P. S. Doyle, *Simulation of electrophoretic stretching of DNA in a microcontraction using an obstacle array for conformational preconditioning*, Biomicrofluidics **3** (2009), 012803, http://bmf.aip.org/resource/1/biomgb/v3/i1/p012803_s1.

[155] C. Valeriani, R. J. Allen, M. J. Morelli, D. Frenkel, and P. R. ten Wolde, *Computing stationary distributions in equilibrium and nonequilibrium systems with forward flux sampling*, J Chem Phys **127** (2007), 114109, http://jcp.aip.org/resource/1/jcpsa6/v127/i11/p114109_s1.

[156] E. Vanden-Eijnden, *Transition path theory*, Computer Simulations in Condensed Matter: from Materials to Chemical Biology (M. Ferrario, G. Ciccotti, and K. Binder, eds.), Lecture note in Physics, Springer, 2006, pp. 453–493.

[157] ———, *Some recent techniques for free energy calculations*, J Comput Chem **30** (2009), no. 11, 1737–1747, http://onlinelibrary.wiley.com/doi/10.1002/jcc.21332/full.

[158] E. Vanden-Eijnden, M. Venturoli, G. Ciccotti, and R. Elber, *On the assumptions underlying milestoning*, J. Chem. Phys. **129** (2008), 174102.

[159] H. Wang, *An improved WPE method for solving discontinuous Fokker-Planck equations*, Int. J. Numer. Anal. Model. **5** (2008), no. 1, 1–23. MR2479115

[160] H. Wang, C. S. Peskin, and T. C. Elston, *A robust numerical algorithm for studying biomolecular transport processes*, J. Theoret. Biol. **221** (2003), no. 4, 491–511, DOI 10.1006/jtbi.2003.3200. MR2043095

[161] N. Woo, E. S. G. Shaqfeh, and B. Khomami, *The effect of confinement on the dynamics and rheology of dilute DNA solutions. i. entropic spring force under confinement and numerical algorithm*, J Rheol **48** (2004), 281, http://journalofrheology.org/resource/1/jorhd2/v48/i2/p281_s1.

[162] ———, *The effect of confinement on the dynamics and rheology of dilute DNA solutions. ii. effective rheology and single chain dynamics*, J Rheol **48** (2004), 299, http://journalofrheology.org/resource/1/jorhd2/v48/i2/p299_s1.

[163] O. Zeitouni, *Random walks in random environment*, Lectures on probability theory and statistics, Lecture Notes in Math., vol. 1837, Springer, Berlin, 2004, pp. 189–312, DOI 10.1007/978-3-540-39874-5_2. MR2071631

[164] Y. Zhang, J. J. de Pablo, and M. D. Graham, *An immersed boundary method for Brownian dynamics simulation of polymers in complex geometries: Application to DNA flowing through a nano slit with embedded nanopits*, J Chem Phys **136** (2012), 014901, http://jcp.aip.org/resource/1/jcpsa6/v136/i1/p014901_s1.

Editorial Information

To be published in the *Memoirs*, a paper must be correct, new, nontrivial, and significant. Further, it must be well written and of interest to a substantial number of mathematicians. Piecemeal results, such as an inconclusive step toward an unproved major theorem or a minor variation on a known result, are in general not acceptable for publication.

Papers appearing in *Memoirs* are generally at least 80 and not more than 200 published pages in length. Papers less than 80 or more than 200 published pages require the approval of the Managing Editor of the Transactions/Memoirs Editorial Board. Published pages are the same size as those generated in the style files provided for $\mathcal{A}_{\mathcal{M}}\mathcal{S}$-LaTeX or $\mathcal{A}_{\mathcal{M}}\mathcal{S}$-TeX.

Information on the backlog for this journal can be found on the AMS website starting from `http://www.ams.org/memo`.

A Consent to Publish is required before we can begin processing your paper. After a paper is accepted for publication, the Providence office will send a Consent to Publish and Copyright Agreement to all authors of the paper. By submitting a paper to the *Memoirs*, authors certify that the results have not been submitted to nor are they under consideration for publication by another journal, conference proceedings, or similar publication.

Information for Authors

Memoirs is an author-prepared publication. Once formatted for print and on-line publication, articles will be published as is with the addition of AMS-prepared frontmatter and backmatter. Articles are not copyedited; however, confirmation copy will be sent to the authors.

Initial submission. The AMS uses Centralized Manuscript Processing for initial submissions. Authors should submit a PDF file using the Initial Manuscript Submission form found at `www.ams.org/submission/memo`, or send one copy of the manuscript to the following address: Centralized Manuscript Processing, MEMOIRS OF THE AMS, 201 Charles Street, Providence, RI 02904-2294 USA. If a paper copy is being forwarded to the AMS, indicate that it is for *Memoirs* and include the name of the corresponding author, contact information such as email address or mailing address, and the name of an appropriate Editor to review the paper (see the list of Editors below).

The paper must contain a *descriptive title* and an *abstract* that summarizes the article in language suitable for workers in the general field (algebra, analysis, etc.). The *descriptive title* should be short, but informative; useless or vague phrases such as "some remarks about" or "concerning" should be avoided. The *abstract* should be at least one complete sentence, and at most 300 words. Included with the footnotes to the paper should be the 2010 *Mathematics Subject Classification* representing the primary and secondary subjects of the article. The classifications are accessible from `www.ams.org/msc/`. The Mathematics Subject Classification footnote may be followed by a list of *key words and phrases* describing the subject matter of the article and taken from it. Journal abbreviations used in bibliographies are listed in the latest *Mathematical Reviews* annual index. The series abbreviations are also accessible from `www.ams.org/msnhtml/serials.pdf`. To help in preparing and verifying references, the AMS offers MR Lookup, a Reference Tool for Linking, at `www.ams.org/mrlookup/`.

Electronically prepared manuscripts. The AMS encourages electronically prepared manuscripts, with a strong preference for $\mathcal{A}_{\mathcal{M}}\mathcal{S}$-LaTeX. To this end, the Society has prepared $\mathcal{A}_{\mathcal{M}}\mathcal{S}$-LaTeX author packages for each AMS publication. Author packages include instructions for preparing electronic manuscripts, samples, and a style file that generates the particular design specifications of that publication series. Though $\mathcal{A}_{\mathcal{M}}\mathcal{S}$-LaTeX is the highly preferred format of TeX, author packages are also available in $\mathcal{A}_{\mathcal{M}}\mathcal{S}$-TeX.

Authors may retrieve an author package for *Memoirs of the AMS* from `www.ams.org/journals/memo/memoauthorpac.html`. The *AMS Author Handbook* is available in PDF format from the author package link. The author package can also be obtained free of charge by sending email to `tech-support@ams.org` or from the Publication Division, American Mathematical Society, 201 Charles St., Providence, RI 02904-2294, USA. When requesting an author package, please specify AMS-LaTeX or AMS-TeX and the publication in which your paper will appear. Please be sure to include your complete mailing address.

After acceptance. The source files for the final version of the electronic manuscript should be sent to the Providence office immediately after the paper has been accepted for publication. The author should also submit a PDF of the final version of the paper to the editor, who will forward a copy to the Providence office.

Accepted electronically prepared files can be submitted via the web at `www.ams.org/submit-book-journal/`, sent via FTP, or sent on CD to the Electronic Prepress Department, American Mathematical Society, 201 Charles Street, Providence, RI 02904-2294 USA. TeX source files and graphic files can be transferred over the Internet by FTP to the Internet node `ftp.ams.org` (130.44.1.100). When sending a manuscript electronically via CD, please be sure to include a message indicating that the paper is for the *Memoirs*.

Electronic graphics. Comprehensive instructions on preparing graphics are available at `www.ams.org/authors/journals.html`. A few of the major requirements are given here.

Submit files for graphics as EPS (Encapsulated PostScript) files. This includes graphics originated via a graphics application as well as scanned photographs or other computer-generated images. If this is not possible, TIFF files are acceptable as long as they can be opened in Adobe Photoshop or Illustrator.

Authors using graphics packages for the creation of electronic art should also avoid the use of any lines thinner than 0.5 points in width. Many graphics packages allow the user to specify a "hairline" for a very thin line. Hairlines often look acceptable when proofed on a typical laser printer. However, when produced on a high-resolution laser imagesetter, hairlines become nearly invisible and will be lost entirely in the final printing process.

Screens should be set to values between 15% and 85%. Screens which fall outside of this range are too light or too dark to print correctly. Variations of screens within a graphic should be no less than 10%.

Any graphics created in color will be rendered in grayscale for the printed version unless color printing is authorized by the Managing Editor and the Publisher. In general, color graphics will appear in color in the online version.

Inquiries. Any inquiries concerning a paper that has been accepted for publication should be sent to `memo-query@ams.org` or directly to the Electronic Prepress Department, American Mathematical Society, 201 Charles St., Providence, RI 02904-2294 USA.

Editors

This journal is designed particularly for long research papers, normally at least 80 pages in length, and groups of cognate papers in pure and applied mathematics. Papers intended for publication in the *Memoirs* should be addressed to one of the following editors. The AMS uses Centralized Manuscript Processing for initial submissions to AMS journals. Authors should follow instructions listed on the Initial Submission page found at www.ams.org/memo/memosubmit.html.

SELECTED PUBLISHED TITLES IN THIS SERIES

1224 **Werner Hoffmann and Satoshi Wakatsuki,** On the Geometric Side of the Arthur Trace Formula for the Symplectic Group of Rank 2, 2018

1223 **Sergey Fomin and Professor Dylan Thurston,** Cluster Algebras and Triangulated Surfaces Part II: Lambda Lengths, 2018

1222 **Maurice Duits and Kurt Johansson,** On Mesoscopic Equilibrium for Linear Statistics in Dyson's Brownian Motion, 2018

1221 **Francesco Lin,** A Morse-Bott Approach to Monopole Floer Homology and the Triangulation Conjecture, 2018

1220 **Paata Ivanisvili, Dmitriy M. Stolyarov, Vasily I. Vasyunin, and Pavel B. Zatitskiy,** Bellman Function for Extremal Problems in BMO II: Evolution, 2018

1219 **Olivier Frécon,** Algebraic $\overline{\mathbb{Q}}$-Groups as Abstract Groups, 2018

1218 **Gabriella Pinzari,** Perihelia Reduction and Global Kolmogorov Tori in the Planetary Problem, 2018

1217 **Chin-Yu Hsiao,** Szegő Kernel Asymptotics for High Power of CR Line Bundles and Kodaira Embedding Theorems on CR Manifolds, 2018

1216 **Robert Lipshitz, Peter Ozsváth, and Dylan P. Thurston,** Bordered Heegaard Floer Homology, 2018

1215 **Lior Fishman, David Simmons, and Mariusz Urbański,** Diophantine Approximation and the Geometry of Limit Sets in Gromov Hyperbolic Metric Spaces, 2018

1214 **David Chataur, Martintxo Saralegi-Aranguren, and Daniel Tanré,** Intersection Cohomology, Simplicial Blow-Up and Rational Homotopy, 2018

1213 **Sebastiano Carpi, Yasuyuki Kawahigashi, Roberto Longo, and Mihály Weiner,** From Vertex Operator Algebras to Conformal Nets and Back, 2018

1212 **Roelof Bruggeman, Youngju Choie, and Nikolaos Diamantis,** Holomorphic Automorphic Forms and Cohomology, 2018

1211 **Shouhei Honda,** Elliptic PDEs on Compact Ricci Limit Spaces and Applications, 2018

1210 **Zhou Gang, Dan Knopf, and Israel Michael Sigal,** Neckpinch Dynamics for Asymmetric Surfaces Evolving by Mean Curvature Flow, 2018

1209 **Cristian Anghel, Iustin Coandă, and Nicolae Manolache,** Globally Generated Vector Bundles with Small c_1 on Projective Spaces, 2018

1208 **Matthew Harrison-Trainor,** Degree Spectra of Relations on a Cone, 2018

1207 **Alastair J. Litterick,** On Non-Generic Finite Subgroups of Exceptional Algebraic Groups, 2018

1206 **Anne-Laure Dalibard and Laure Saint-Raymond,** Mathematical Study of Degenerate Boundary Layers: A Large Scale Ocean Circulation Problem, 2018

1205 **Charles Collot,** Type II Blow Up Manifolds for the Energy Supercritical Semilinear Wave Equation, 2018

1204 **Rui Palma,** Crossed Products by Hecke Pairs, 2018

1203 **Xiao Xiong, Quanhua Xu, and Zhi Yin,** Sobolev, Besov and Triebel-Lizorkin Spaces on Quantum Tori, 2018

1202 **Naiara V. de Paulo and Pedro A. S. Salomão,** Systems of Transversal Sections Near Critical Energy Levels of Hamiltonian Systems in $\mathbb{R}^4$, 2018

1201 **Bernhelm Booß-Bavnbek and Chaofeng Zhu,** The Maslov Index in Symplectic Banach Spaces, 2018

1200 **Francis Nier,** Boundary Conditions and Subelliptic Estimates for Geometric Kramers-Fokker-Planck Operators on Manifolds with Boundaries, 2018

For a complete list of titles in this series, visit the
AMS Bookstore at **www.ams.org/bookstore/memoseries/**.